International Relations and the Philosophy of History

International Relations and the Philosophy of History

A Civilizational Approach

A. Nuri Yurdusev
Associate Professor of International Relations
Middle East Technical University, Ankara, Turkey

palgrave
macmillan

First published 2003 by
PALGRAVE MACMILLAN
Houndmills, Basingstoke, Hampshire RG21 6XS and
175 Fifth Avenue, New York, N.Y. 10010
Companies and representatives throughout the world

PALGRAVE MACMILLAN is the global academic imprint of the Palgrave Macmillan division of St. Martin's Press, LLC and of Palgrave Macmillan Ltd. Macmillan® is a registered trademark in the United States, United Kingdom and other countries. Palgrave is a registered trademark in the European Union and other countries.

ISBN 0–333–71363–X

This book is printed on paper suitable for recycling and made from fully managed and sustained forest sources.

A catalogue record for this book is available from the British Library.

Library of Congress Cataloging-in-Publication Data

Yurdusev, A. Nuri.
International relations and the philosophy of history : a civilizational approach / A. Nuri Yurdusev.
p. cm.
Includes bibliographical references and index.
ISBN 0–333–71363–X (cloth)
1. International relations–Philosophy. 2. History–Philosophy.
3. Historiography. I. Title.

JZ1305.Y87 2003
327.1'01–dc21 2003049805

10 9 8 7 6 5 4 3 2 1
12 11 10 09 08 07 06 05 04 03

Printed and bound in Great Britain by
Antony Rowe Ltd, Chippenham and Eastbourne

Contents

Acknowledgements

In the course of writing this book, the bulk of which draws on my D.Phil thesis, I have been the recipient of the help of many people. I would like to express my gratitude to Murray Forsyth, my doctoral supervisor at the University of Leicester, for his invaluable guidance and suggestions. I am deeply indebted to Geoffrey Berridge, who not only commented on an early exploratory essay which now constitutes some parts of Chapters 1 and 2, but also gave me all the encouragement to publish the work. He has been a real friend in the process leading to the publication of this book. John Hoffman and Costas Constantinou commented on the first draft and strongly encouraged me to go for publication. James Joll, the late Elie Kedourie, J. D. B. Miller, the late Peter Savigear and Kenneth W. Thompson kindly gave their notes on the initial proposal. I should like to record my thanks to all of them.

If ever the standard author's dedication '... without whom this work could not have been completed' applied, it does to Esin, my wife, whose immense support and warmth has made it all bearable.

Introduction

'A story has no beginning or end: arbitrarily one chooses that moment of experience from which to look back or from which to look ahead.' Such are the words Graham Greene wrote in the very beginning of one of his so-called 'catholic' novels, *The End of the Affair*.[1] It is very easy to dismiss these remarks as the utterances of a 'cynical catholic' writer. Yet, they can be considered, as I do consider them, to be the expression of the obvious uncertainty and, paradoxically, immediate certainty of the process which we, as human beings, all go through and experience. Uncertainty, because the point from which we look back or ahead is not just a point, but one that has backward and forward extensions, extensions to which we have no fixed limits. Certainty, because there is a point in which we find ourselves, not 'arbitrarily chosen' as Graham Greene knew it, only too well.

We find ourselves in a place. Because we have to be somewhere, here or there. When we come to know it we become aware of the fact that the place in which we find ourselves is not a simple 'here', rather a 'here' distinct from a 'there'. When we pursue it further we recognize other 'heres' or 'theres' and we see that heres and theres are interchangeable, a 'here' for us is a 'there' for someone else and a 'here' for now may become a 'there' later. So, we experience uncertainty in the innumerable heres and theres, and certainty in the particular here in which we find ourselves, or in which we have to be. However when we think of the totality of all heres and theres, we get the idea of '*space*', named in its various extensions and comprehensions such as 'region', 'locality', 'country', 'territory', 'island', 'continent', the 'Earth', 'galaxy', the 'Universe' and so on.

We find ourselves in a place. While we stand in the place, we witness that it changes together with other places. The sun rises, the day dies, and flowers come out, leaves fall down. We experience and go through the 'occurrences' and 'happenings', 'differences' and 'variations', and 'changes' and 'alterations'. Our experience of them leads us to the idea of *'time'*, accordingly named in its diverse extensions and experiences such as 'hour', 'day', 'week', 'month', 'year', 'century', 'age', 'millennium' and so forth. We again see uncertainty and certainty. Uncertainty, because the moment we instantly experience, 'present', has no duration; the moment we experienced, 'past', has ceased to exist; and the moment we shall experience, 'future', does not exist yet. Certainty, because we experience it. Moreover, the uncertainty described in such terms, paradoxically, turns into a certainty – the continuity of time, or process of experience.[2]

We find ourselves in a place. If we look around, we come across other men in the same place. Other men finding themselves in the same place and, perhaps, experiencing the same occurrences. Then, we understand that we are not single individuals, living and experiencing alone. More significant still, there are those that have already been there and been experiencing. We come to realize that we do not only share the place and time, but also the way we experience, the way we become aware of things. Consequently, we reach the idea of *'society'* or *'men in society'*, similarly named in its different extensions and sharings such as 'society', 'group', 'community', 'tribe', 'clan', 'nation', 'civilization', 'humanity' and so on. The same uncertainty and certainty exist. Uncertainty due to the fact that we find ourselves together with many others. Certainty due to the fact that we always find ourselves with some others.

We find ourselves in a place, at a particular point in time, and with others (or within a society). When we reflect upon space, time and society, we get *'knowledge'*, named again in terms of the contents and extensions of the reflection such as 'science', 'theory', 'history', 'art', 'literature' and so forth. Knowledge and reflection could further be sub-categorized as 'physics', 'chemistry', 'astronomy', 'archaeology', 'sociology', 'economics', 'politics', 'international relations' and so on. This book, which is a reflection on the triad of space, time and men in society is intended to be within 'international relations' or for those interested in what is called 'international theory',[3] because of the interests and accumulation of its author, or due to the place where the author finds himself. However its scope far extends the established limits of International Relations. It combines basically International Theory, Political Theory, Sociology, History and Philosophy: naturally, I could say.

More specifically, the present essay purports to examine the links between the study of International Relations and the Philosophy of History, taking the concepts of 'universal history', 'civilization' and 'modern international system' to be the main points of focus. Civilizations and international systems are to be understood as societies/ social entities comprising certain aspects of man's life in space, time and society. Stated in another way, they are creations of men at some point in time and space. The book basically argues that civilization and international system can rightly be considered as units of analysis in international relations, or for the students of International Relations.

In what follows, the stage is opened by chapters on 'theory' and 'history', dealing with such concepts as theory, knowledge, science and history. Relying on the accounts by Collingwood and Oakeshott of knowledge, and following the Kuhnian conception of science in paradigmatic, communal and consensual terms, Chapters 1 and 2 propose a historical understanding of knowledge, science, and history as against the age-old dichotomy between 'objectivism' and 'subjectivism' (or relativism). The basic consideration has been the view that without having an idea on those fundamental issues, it is hardly possible to explicate civilizations and international systems. The stage is completed by an elucidation of the concept of universal history (Chapter 3). Conceived as embracing the totality of all known spatial, temporal and social extensions, universal history is thought to form an adequate basis for the study of civilization and international system.

The stage having being set, the remainder of the present essay is devoted to the comprehensive examination of the concepts of civilization and the modern international system. First, the concept of civilization (Chapters 4 and 5) has been examined in its various meanings and civilization is designated as a large-scale social unit, a unit closely related to international system, for most civilizations, historically, comprised international systems. Starting from this observation and pointing out that international systems as large-scale social units, like civilizations, could be (and have been) either uni-civilizational or multi-civilizational; the modern international system is, then, to be analysed in Chapters 6 and 7 stressing that it is an international system, composed of nation-states and involving multiple civilizations, at least, for the present century. The book concludes that a civilizational analysis is still relevant to the study of international relations, despite the fact that the modern international system is principally composed of nation-states and that it is now world-wide.

1
Theory or Coffee without Sugar

With a curious statement in the first of his essays collected in *Civilization on Trial*, Toynbee tells us how he came to be a historian: 'Why am I a historian, not a philosopher or physicist? For the same reason that I drink tea and coffee without sugar.' As if he would like to reply to our wonder about what sort of a relationship could possibly be found between having tea or coffee without sugar and being a historian, Toynbee adds: 'Both habits were formed at a tender age by following a lead from my mother. I am a historian because my mother was one before me.'[1] The passage could be viewed as showing how effective a mother can be on her infant, or indicating that a child's upbringing or family environment could have a lasting impact upon his adult life. It can also be considered to be an expression of that ever-existing question of whether an individual person makes his own decisions, acts in the way he chooses and does whatever he wishes by his own will; or whether his thoughts, desires, aims, actions and deeds are formed and shaped by his society: in other words, the problem of social and environmental determination versus individual free will. It may be objected that the question is not necessarily to be formulated at such extremes, that the individual and his social environment are not to be taken as in fiercely opposite terms, and that the case is not a one-way determinate relationship. Rightly so. Since Aristotle, we know historically and for certain that the individual person, who is born into a society, grows up in a society and lives within a society, is influenced and moulded by his social surroundings. In the case of Toynbee, by his self-admission, this is the case: perhaps, in a greater degree.

Followed further, other questions arise. Does the social environment of a human being affect only his decisions and desires, actions and deeds and make him, say, a historian? Or, is the way he decides and acts also under the influence of the same environment? In other words, having become a historian, does Toynbee make his research and perform the activity of being a historian according to a set of identifiable and applicable principles or rules outside his particular society or community of fellow historians; or rather, does he carry it out, again, within the bounds of a particular time and place? Toynbee's answer to the second part of the question is affirmative: 'The human observer has to take his bearings from the point in space and moment in time at which he finds himself; and he is bound to be self-centered; for this is part of the price of being a living creature.'[2]

Not only the lives of human beings take place in time and place, but also the way they know, think, reflect and theorize about their lives is also time- and space-bound. Hence, theory versus history – theory as the account of what happened, history as what actually happened; theory as what is said to exist, history as what does exist. This chapter will, therefore, first address the problem of theory (knowledge) and, then, a particular branch of theory (knowledge), that is, science in the Kuhnian understanding, is to be analysed.

Knowing and thinking

When I say 'I know something', it means that I am aware of it; I understand it and I can comprehend it. There is a relation of familiarity between what I do know and myself. It is internalized by me and no longer outside my realm of knowledge. To know thus means to understand, to comprehend and to become aware of the reality with which I and others are faced and in which we all are embedded. Knowledge/theory[3] is, then, a combination of the knower/theorizer and the known/theorized. The knower is I, and the known is what I face, the reality including myself. It follows that when knowing/theorizing, the knower/theorizer does not operate in and by his own existence (domain, rules, system) independent from the known (the theorized); instead he carries out his activity together with and in the domain of the known (the theorized). The knower is thus engaged in a web of relations composed of his personal dispositions and those of the known.

How exactly does knowing occur and what constitutes knowledge? Among other things such as intuition and senses, the process of knowing emerges out of the act of thinking which is said to be the prerogative of man in the world.[4] This remark is, however, far from being explanatory for it itself begs explanation. The questions 'How does knowing occur exactly?' and 'What constitutes knowledge?' apply to the act of thinking as well: how does thinking occur and what constitutes thought?

As an ability or feature of man, the act of thinking could be either caused by something else, say, mind/intellect, or it could be self-creative. Suppose it is caused by mind, as it was supposed by Descartes and many other rationalists. In this view, it is mind which performs the act of thinking, thinking of ideas, producing notions, making distinctions between the true and false, meaningful and meaningless, right and wrong, sensible and insensible, reasonable and unreasonable. The question asked above, again, applies to the mind. In response to the question of how the mind does its act of thinking, two answers may be advanced: by a creator, or by itself. If it were attributed to something beyond mind such as God or Nature, then, the logical question would be what/who gives God or Nature the power to make (man's) mind think. And as an answer, it can only be said 'by itself'. If the ability or power, which makes thinking possible, is said to be by mind's own nature, then, the authority given to God or Nature is this time given to the mind. The two so-called different views then converge in attribution of omnipotency, in one case to God or Nature and in the other to mind itself. Both can rightly be named as rationalist – for God or Nature of the first view is understood to be a reasoning, perhaps all-reasoning, being and, in the second case, the mind itself is but reasoning. They further converge in taking mind, whether it gets its ability of thinking from a being beyond or from itself, as an apparatus to think, an apparatus to know and understand. In other words, the rationalist idea is that you have first a mind and then you think through it. This view of mind as an apparatus for thinking is, to Oakeshott, the basic error of rationalism.[5]

Let us take the view that considers the act of thinking to be self-creative. It would perhaps be better to follow it through its two most prominent advocates: Collingwood and Oakeshott. Not surprisingly, in this view, there is no such thing as the originator of thinking, for 'the activity of thought is a free or self-creative process, which depends on nothing else except itself in order to exist'. Existence is in fact the activity of thinking. Thinking occurs when thinking. What is called thinking is thus the activity of becoming aware of the flow or continuity of sensations and feelings and of the succession of experiences.[6] It is not a mere state of consciousness, it is self-consciousness. Self-consciousness is not, however, a self-activity or a self-possession. Collingwood emphasizes the social and historical nature of thinking: 'The body of human thought or mental activity is a corporate possession, almost all the operations which our minds perform are operations which we learned to perform from others who have performed them already.'[7]

In this view, the idea of mind is not excluded from thinking which is a self-creative activity, but the idea of mind as an apparatus has no place in either Collingwood's nor Oakeshott's view. For them, mind is not a capacity acquiring thoughts and ideas. It is not the creator of thought or thinking. Simply, in Collingwood's words, 'mind is what it does' and 'a mind is nothing except its own activities'.[8] Oakeshott agrees entirely. Mind is the 'offspring of knowledge and activity'. It is not an apparatus first acquiring a filling of ideas and then making distinctions such as true and false, right and wrong, reasonable and unreasonable and so on. 'Properly speaking', declares Oakeshott, 'the mind has no existence apart from, or in advance of, these and other distinctions. These and other distinctions are not acquisitions; they are constitutive of mind.'[9] Mind is part of the thinking process. Thinking goes on within a web of social relations, through other thinking beings. Thus, thinking becomes a socio-historical activity. This view is, I think, supported by the case of the human infant who does not perform the activity of thinking by himself, at least, until a certain age, and with whom we do not associate those characteristics supposed to be of thinking/knowing, characteristics we normally attribute to adult persons. The socio-historical character of thinking, and thus knowing, inevitably leads to the question of, to use the familiar terms, objectivity and subjectivity.

Objectivism versus subjectivism

In its positivistic formulation that has been very influential in the modern period, from the eighteenth century onwards at least, the idea of objectivity, simply stated, refers to the view that knowledge should be attainable and applicable by anyone, anywhere and at any time. Objective knowledge is attainable and applicable by every analyst or scientist because it is intersubjectively certifiable, 'empirically testable' and independent of individual opinion.[10] The principle of objectivity or objective knowledge is associated with the experimental data or experimental observation. This is in line with the positivistic separation of experiment/experience/observation/practice from theory/hypothesis/theoretical statements, a duality whereby the former term is conceived to be objective data. It is thought to follow empirical phenomena through the sensual perception and to form the ground for objective knowledge. The latter term is regarded as the result of mere reasoning, an occurrence which could only lead to mere conviction.

The roots of the positivistic separation in which one element is taken as essential and real and the other is seen to be superfluous and many other such binary oppositions can, of course, be traced back to the ancient periods. It appears in Plato's writings, most famously in his separation of the 'realm of ideas', which is 'the real' world and the 'reality' itself, from the 'realm of appearances' which is no more than a merely seeming reality. Russell makes the point that the distinction was first set forth by Parmenides. One can even take it back to Zarathustra's doctrine of 'last things'. Kant is, perhaps, the most zealous advocate and Nietzsche is the most ardent opponent of these binary oppositions in the modern period.[11]

The modern positivists sustained this dual conception of the reality and knowledge which can, they assumed, only be obtained via sensual observation and experience. For Hempel, empirical knowledge comes from experimental findings. Nagel agrees with him in marking the basic difference between experimental laws and theoretical laws as being the labeling of the former as inductive generalizations based on relations found in observed data.[12] In a sense, objective knowledge is not, for the positivist, time-and-space bound, for not only are the empirical data assumed to have a quality available to anyone, but also all the observers are considered to be able to detect that quality. At the core of the principle of objectivity is the notion that the knower is detachable from the known. Behind this conception of objectivity as such lies what one critic called the 'assumption of objectivism' – 'the basic conviction that there is or must be some permanent, ahistorical matrix or framework to which we can ultimately appeal in determining the nature of rationality, knowledge, truth, reality, goodness, or rightness'.[13] Conceived this way, objectivity implies universality by definition. In fact, objectivity is nothing but universalization of sameness. Objectivity thus becomes universalism.

It is my contention that objectivism does not stand logical and empirical scrutiny. Objectivism, as I have already said, involves some assumptions about the knower and the known. The known is assumed to be accessible in its uniformity to the knower. In other words the known is abstracted from its spatio-temporal domain. The process of change is excluded from the analysis. We know from everyday life that this is not the case. Otherwise, it would be a denial of all those notions on which our life is based, such as alternation of day and night, and the shift in, say, climatic and geographical occurrences. Furthermore, as the cliché goes, facts do not speak themselves, implying the impossibility of pure empiricism. True, in nature, events recur; there is the repeatability of events and facts. So, facts may speak for themselves, but only with *ceteris paribus* clauses. Then, repeatability does not imply an unchanging exactness and uniformity. It could, further, be argued that 'all facts are unique, those dealt with by natural scientists no less than any others'.[14] Even if facts in nature are deemed to have uniform characteristics, it is not the case that they are awaiting discovery in themselves. They are discoverable, but as already mentioned, with their accompanying conditions. Water boils at 100°C, but only at sea level, not at the peak of the Himalayas. The facts are then discoverable not in themselves but with others, and they do not speak for themselves, but with *ceteris paribus* clauses, that is, with those who examine them. It follows that no fact can lead to objectivism. Here, we come to the other foot of the bridge – the knower.

On this side, too, objectivism does not stand scrutiny. Here, it is based on two assumptions depending upon those related to the other foot – the known. Assuming that the known is the same, that is to say, that facts are uniform, objectivism supposes a knower who has no interest, that is to say, a disinterested observer. This again means the denial of all that we associate with being a human being, an active and willing being. It is nothing but a conceptual fiction to posit a 'pure, will-less, painless, timeless knowing subject' and 'contemplation without interest' is a 'nonsensical absurdity'.[15] The mind of the analyst (a knower and however his or her mind may be defined) before his or her object (a known and whatever that may be) is hardly an empty bowl. It brings with it its own set of ideas. The knower never simply surrenders to the known. The corollary is that, in the words of White, 'pure interpretation, the disinterested inquiry into anything whatsoever, is unthinkable'. Inquiry of any sort is unimaginable without some kind of presupposition.[16] The knower does not passively respond to the impulses that emanate from the so-called external world. Even if mind is defined in instrumental terms, thinking, knowing, understanding, all these cognitive processes do not occur in a vacuum. On the contrary, they all require some tools for their functioning, such as concepts and words. Gadamer is right, in declaring that 'all understanding [knowing] inevitably involves some prejudice'.[17] Pure empiricism once more falls beyond the range of feasibility in the human world.

The other assumption objectivism posits with respect to the knower, in case the uniformity or sameness does not apply to the known, envisages the attribution of sameness or a universal characteristic to the knower. If, say, the knower has a mind, it is assumed that it is universal in the sense that everyone (all knowers) has the same ability or apparatus. Since everyone has the same mind or thinking process and acts in the same rationality, then, everybody detects the same knowledge out of the known no matter how variously patterned it is. Even if we accept such an assumption about the existence of a mind or rationality potentially existing in all human beings, that is no guarantee for objectivism. The objectivist needs further guarantees for consolation. The tools the thinking mind uses or the environment in which the thinking process takes place should not vary. However, we do not have a universal language. There is no evidence, to use Popper's term, to falsify Winch's argument that it is in principle impossible that a human being should learn a language outside a human society.[18] If man does not (and cannot) learn a language outside a society, it means he cannot engage in knowing/theorizing without a society, for the medium of this activity is socially based.

Following Collingwood's and Oakeshott's view that the mind is part of thinking and thinking is a 'corporate possession', we could go further and assert that not only the medium, but also the activity of thinking itself is socio-historical. This is the logical and, one could say, empirical outcome of man being a social creature. Man becomes a man out of merely being within a society. Since there is not a monopoly of one single socio-historical environment in the world so far as we know it, then, it goes that the activity of knowing could be performed and expressed differently according to different socio-historical environments. This gives way not to the objective knowledge of objectivism, but to the knowledge supplied by the objective conditions (social and historical conditions) of the knower and, in some cases such as in the fields of humanities, of the known. Even though the act of thinking as man's distinctive feature can be taken as having universality, the way it is performed and expressed is always particular. 'Universal thought', said Donelan, 'needs local dress.'[19] Both universalism and pure empiricism, therefore, have no validity in the empirical world in which we live. Objectivism demands far too many assumptions to be satisfied.

That objectivism, as defined here, does not hold for human knowledge is clear enough. Does the alternative, the idea of subjectivism or what Bernstein calls 'relativism', apply? Relativism is defined as 'the basic conviction that when we turn to examination of those concepts that philosophers have taken to be the most fundamental – whether it is the concept of rationality, truth, reality, right, the good or norm – we are forced to recognize that in the final analysis all such concepts must be understood as relative to a specific conceptual scheme, theoretical framework, paradigm, form of life, society or culture'.[20] Relativism or subjectivism seems the exact opposite of objectivism: what is not objective is subjective, what is not universal is relative or local.

For objectivism, there is a framework or a universal rationality, but for subjectivism there is no such thing.

Just as the roots of objectivism can be traced back to the Greeks, one could find the traces of subjectivism/relativism in Greek philosophy, basically in the Sophists. One of the Sophists, Gorgias, maintained that nothing exists; that if anything exists, it is unknowable; and granting it even to exist and to be knowable by any one man, he could never communicate it to others.[21] In our time Feyerabend's 'anarchistic theory of knowledge' is perhaps the typical expression of relativism. Here, we are given an endless and limitless plurality of theories. 'There is', declares Feyerabend, 'only *one* principle that can be defended under *all* stages of human development. It is the principle: anything goes.'[22] No principle and no theory are commonly applicable or valid for anything can be developed and anything goes and every theory applies. It is a vain attempt to propound and develop commonalities. For the relativist, in its extreme case, theory or knowledge is anything you can get away with.

Stated in these terms, relativism, just as its opponent, objectivism, depends upon a couple of assumptions that cannot be maintained logically and empirically. First and foremost, there is the fallacy of relativism in which relativism renders itself relative and is thus logically self-refuting. Leaving aside this fallacy, the basic assumption of relativism could be expressed as follows. First, every individual knower is ascribed with an absolute competency by which he could do and develop whatever he wants. While objectivism reifies singleness or oneness, relativism reifies plurality and multiplicity. In the case of objectivism one stance/theory is universalized and others are denied; in the case of relativism, all theories are accepted and every theory is in a sense universalized. Such a conception makes man too indeterminate and denies the very basis of society and thus man himself.[23]

On the part of the known, relativism, followed to its own logical consequences, requires unconnected, clearly separated facts. However, we know that there are regularities and patterns in both nature and society even if with accompanying conditions, which make it difficult to hold the view that 'anything goes'. This remark does not, however, deny that 'some things go'.

Second, relativism depicts a universe of knowers, and in fact knowns, that are fixed, static, closed and incapable of communicating to and interacting with each other. It means that men and societies are divided by unbridgeable differences. But our knowledge and understanding of men and other societies depends, to some extent, on the possibility of understanding others and the existence of communication with others. Furthermore the denial of some common values would mean the denial of all that we associate with human morality.[24] The existence of commonalities is a pre-condition of man's becoming man. Relativism, as already stated, assumes that men (the knowers) and their socio-historical environments (the knowns, plus what makes men know) are static, closed, fixed and free from interaction and communication with others. However, this assumption is rejected by the idea of change. We know historically that societies do not exist at the present time as they did at the dawn of time. Therefore, 'the very existence of change in history directs us to the possibility of understanding other traditions because it denies the "fixedness" or "closedness" that relativists presuppose or imply'.[25] Absolute relativism, just as objectivism, fails in the face of the empirical/historical world. Both of them are based on reifications: one reifies the general (objectivism) and the other reifies the particular (relativism).

Historical conception of science and knowledge

Contrary to the reifications of objectivism and relativism, Thomas S. Kuhn provides us with a historical conception of science (and thus knowledge) by examining the history of science. Kuhn's paradigmatic view of science[26] may, I think, be taken as a ground for arriving at a historical conception of knowledge. A few remarks follow on why science and why Kuhn seem to be necessary.

Of the branches of knowledge such as science, history, theology, mathematics and, possibly, arts; science[27] has had and still has an enormous preeminence and pertaining power not only in Western culture but in other cultures too, at least in the modern era. Science is not just an intellectual endeavour. It has a great impact in the everyday life of human beings, thanks perhaps to the traditional, and not necessarily correct, association of the enterprise of science with technological advancements. Kuhn, as a historian of science, tells us that until late in the nineteenth century there had not been a significant association of technological innovations with the men, the institutions, or the social groups that contributed to the science.[28]

Discussions on how scientific so-called social sciences are and attempts to make social disciplines as 'true' sciences as natural sciences simply show the influence of science on the students of various social disciplines or subjects. The issue of scientificity has been the focal point in social fields ever since the overwhelming success and prestige of natural sciences dating from the seventeenth and eighteenth centuries. The subjects of international relations and history are no exceptions. To cite some examples of discussions on the issue in our time: Nagel argued that the scientific social science was possible in spite of the difficulties in establishing general laws in social studies, for the difficulties could only present practical challenges, not conceptual impossibilities. Winch, however, disagrees with Nagel, arguing that what is required to explain the human society is *logically incompatible* with the explanation offered in natural sciences, that the understanding of society is *logically different* from that of nature, and that central concepts for understanding social life are incompatible with those central to scientific prediction. Popper and Giddens agree with Winch on the qualitative differences between subjects of society and nature.[29] For international relations, we need only to recapture the well-known 'second debate'.[30]

As to 'why choose Kuhn?', leaving aside the impact of his paradigmatic view for science in paving the way for so-called 'post-positivist' philosophies of science and for the discussions on scientificity and nature of social disciplines,[31] I think, his conception of science as a communal endeavour and his view that science is practised on a consensual basis entail a historical and common-sense understanding of science and thus knowledge. A brief review of his paradigmatic view of science would be useful here.

Kuhn examines the history of science and focuses on its social and communal aspects, instead of formulating abstract rules for scientific activity as the positivists and Popperians do. To understand the nature of science, its method, technique, theories and practices, we need to study the community structure of science and the way scientists carry out their activities. In other words, Kuhn has rendered science a 'practice of scientists', rather than a universally applicable credo, detachable from the individual scientist who is supposed to have no association with the framework in, and by, which the community operates, save blind obedience. Science then becomes what the community of scientists does. Kuhn expresses it in an acute style at the very end of the 'postscript 1969': 'scientific knowledge, like language, is intrinsically the common property of a group or else nothing at all. To understand it we shall need to know the special characteristics of the groups that create and use it.'[32] Then, what is a scientific community?

If science is what the community of scientists does, then, is it the case that a scientific community is composed of those who say that they are scientists and who identify themselves with the community? Exactly so. The identification of scientific communities is based upon the existence of paradigms and vice versa. 'A paradigm is what the members of a scientific community share, conversely, a scientific community consists of men who share a paradigm.'[33] For Kuhn, this is not a vicious circle, though it presents difficulties. Scientific communities can, however, be insulated without recourse to paradigms. First and foremost, scientists of a community have undergone similar education and professional instructions. They have absorbed the same technical literature and they see themselves, and are seen by others, as responsible for the pursuit of a set of shared goals and principles marking the limits of their subject matter. There is a relatively full communication and also a relatively unanimous judgment.[34] In sum, a scientific community is created as a result of a socialization process and mutual recognition among the present and/or prospective members.

Having defined the scientific community, next comes the concept of paradigm. A paradigm is simply what the members of a scientific community share. In the 'postscript 1969', Kuhn elucidated the meaning of the concept. He tells us that the term paradigm is used in two different senses in the original book. 'On the one hand, it stands for the entire constellation of beliefs, values, techniques, and so on shared by the members of a given community. On the other, it denotes one sort of element in that constellation, the concrete puzzle-solutions which, employed as models or examples, can replace explicit rules as a basis for the solution of the remaining puzzles of normal science.'[35] A paradigm then, in its broad definition, involves the entire constellation of assumptions, values, standards, agendas, models and programmes shared by the scientific community. A paradigm is accordingly a research framework that guides a scientific community and under which normal scientific activity takes place.

The implications of the concept of paradigm as the entire constellation of beliefs and values for science are immense. First, it means that we have no way of viewing the facts in themselves and evaluating rationally the theories or researches (activities of the scientists). Second, it implies that scientists do not and cannot emulate a positivistic or Popperian account of their practice – testing empirically with a universally applicable method. A universal empirical testability and rationality cannot be maintained because our access to the facts through which we test our beliefs (theories) are always filtered by our existing paradigm. Kuhn makes this point clear when arguing for 'the priority of paradigms'. Paradigms are 'prior to, more binding, and more complete than any set of rules for research'.[36] The conception of science as paradigm-governed and the term paradigm as a constellation of beliefs, values, models, all that a community shares, strip science of its universality, that is to say its proclaimed validity and truth for all (rational) persons. Such a conception of science – paradigm-governed and community-based – implies the possibility of multiplicity in the understanding and practice of science. What it also entails is the concept of the changeability of one particular understanding and practice in the course of history. To paraphrase the earlier statement on knowledge, science is, in this view, a combination of the scientific community, which is marked by a guiding paradigm, and the subject or facts the community examines.

Being defined as a paradigmatic and communal endeavour, scientific activity, according to Kuhn's analysis of its history, goes on at three distinct phases: normal science, period of anomaly (or crisis), and scientific revolution. Normal science is the activity of the scientific community within a research framework, or guiding paradigm. Kuhn defines 'normal science' as 'research firmly based upon one or more past scientific achievements, achievements that some particular scientific community acknowledges for a time as supplying the foundation for its further practice'.[37] The most striking feature of the research activities under normal science is that they seldom aim to produce novelties. The paradigm provides the community with a criterion to choose problems that can be assumed to have solutions. Most of the scientific work during the period of normal science seeks to solve puzzles within the framework of the prevalent paradigm. They may extend or articulate the paradigm over time.[38] Normal science then consists of routine operations by a community of scientists/practitioners trying to solve puzzles under the guidance of an agreed paradigm framework, one through which, at once, individual practitioners are socialized. Normal science is a cumulative enterprise and the continuity of this routine activity is disturbed only with the emergence of anomalies.

After normal science comes a period of anomaly or crisis. Anomaly is the findings that do not conform to the prevalent paradigm, in Kuhn's words, 'the recognition that nature has somehow violated the paradigm-induced expectations that govern normal science'. The first response of scientists in facing anomalies is to try to adjust the contrary findings to the theory. Even if they may begin to lose the faith in the existing paradigm they do not yet renounce it. When anomalies within the paradigm become increasingly severe, in other words, when they come to seem more than just another puzzle of normal science, the transition to crisis or extraordinary science has begun. The paradigm in crisis is rejected only if there is an alternative formation. This means that in periods of crisis, competing paradigms emerge. The decision to reject one paradigm and accept another is taken by a comparison of both paradigms with nature and also with each other.[39] The transition from one paradigm to another (paradigm shift) is what Kuhn calls a 'scientific revolution'.

A scientific revolution, out of which a new tradition of normal science emerges, is far from a cumulative process. It involves a reconstruction of the field with new fundamentals, changing the theoretical generalizations, methods, models and applications. It starts with a growing sense and awareness, usually in a subdivision of the scientific community, that the existing paradigm has ceased to function adequately in the explanation of one particular aspect of nature. Scientific revolutions resemble political revolutions. In Kuhn's bold remarks: 'Like the choice between competing political institutions, that between competing paradigms proves to be a choice between incompatible models of community life ... As in political revolutions so in paradigm choice – there is no standard higher than the assent of the relevant community.' The decision is made, not on past achievement as the old paradigm has already failed, but on future promise, and such a decision 'can only be made on faith'.[40] In sum, in the course of a scientific revolution, there may be (and are) different communities working simultaneously within different paradigm-governed frameworks, each being incommensurable (they cannot always be measured against each other), and incompatible (meaning that they are in conflict with each other). They are, however, comparable with each other, without requiring a fixed common assumption. The choice between competing paradigms cannot be made on the basis of an objective or neutral criterion, but on the basis of group discussion and debate, a socio-psychological and historical process. In plain words, it is a matter of which one of the contending paradigms can persuade and gain enough converts to become the dominant framework for research and scientific activity. For Kuhn, scientific revolutions are rare occurrences in the history of science such as the shift from Newtonian to Einsteinian physics. Once this has happened, the new period of normal science begins.

Such is Kuhn's paradigmatic conception of science. Turning back to the initial question of this chapter – objectivism versus relativism – where does Kuhn stand? It seems quite obvious that he is not within the lands of objectivism, for his paradigm depends upon a social and historical community and, furthermore, the formation of paradigm, its components, beliefs constituting it, and the choice about it are not wholly based on what may be termed 'objective' tools such as observation, experience and a common rationality. As he expresses it clearly, there is always an arbitrary element, compounded of personal and historical accident, as a formative element in what a scientific community shares, that is, paradigm.[41] The objectivist notions such as being independent of time and space, pure empiricism, determination of theory by data, absence of any metaphysical, theological and arbitrary factors in the assessment of theories and methods, neutral observation, universal rationality, comparison of the theories only with bare nature and empirical accuracy as the sole criterion for the acceptability of theories do not seem to have a place in Kuhn's understanding of science.

Kuhn's paradigmatic account of science is not objectivist. If so, then, is it relativist? One's first impression of *The Structure of Scientific Revolutions* is that its basic argument is relativistic as it renders science a relative enterprise, relative to a specific community.[42] Given the concept of paradigm as a constellation of beliefs by a community, containing irrational and arbitrary elements and being admitted largely as a result of its success, and his view that paradigms are incommensurable and the members of different communities live in 'different worlds', what else one can make out of it but that relativism is a legitimate and right conclusion.

Kuhn admits this charge of relativism, but it is not, he asserts, a mere relativism. He describes the development of the modern scientific specialties similar to a tree with roots, trunk and branches. Among the branches, he says, common criteria are possible as they are all linked to the same trunk.[43] Kuhn implies that, once a paradigm is established, an objective criterion for research can be proposed. In addition, Kuhn says that research, in science, is carried out under a single paradigm during long periods of time (during the period of normal science). This is indeed the characteristic that gives science its uniqueness, for no other activity has been carried out for long periods under one single framework.[44] Applied to scientific activities within a paradigm, and considering his historical observations that in normal periods usually one single paradigm prevails, Kuhn's position is more than simply relativistic. A common criterion is attainable not by a universally applicable definition, but by a socially agreed system of principles and rules. This basic point has been further expressed in his treatment of contemporary social sciences.

The present day social sciences are, for Kuhn, at the pre-paradigm stage, where the discussions about whether they are really a science are abundant. These concerns for scientificity in social sciences will cease to appear 'not when a definition is found, but when the groups that now doubt their own status achieve consensus about their past and present accomplishments'.[45] The significance of Kuhn's work is that it has shown us that such consensus is attainable in practice as achieved in the natural sciences. The consensus need not be in every minute point. Nothing in his work, to my understanding, adumbrates a position like this. To sum up, Kuhn's conception of science does not seem to rely on 'objectivism' and 'subjectivism', in the sense of this essay. True, it is relativistic to a degree, when it comes to 'fundamentals'. It is not, however, relativistic in practice, as some degree of 'historical objectivity', whose standards depend on the existence of a community of inquirers who are engaged in argumentation and in consensus, does exist. Not everything goes at any time.

As already pointed out, science is but one branch of knowledge, in other words, scientific knowledge does not exhaust the totality of human knowledge. The knower of the scientific knowledge is in general human beings just the knower of all knowledge, in particular the community of those who are professionally engaged in the activity of knowing and producing the scientific knowledge, that is, scientists. The known of scientific knowledge is nature, the physical world we perceive. Scientific knowledge is then a combination of nature and the community of scientists. One would say that science emerges out of the interaction between the community of scientists and their object, that is, nature and its events. It does not only stem from nature, natural events, their occurrences, changes, recurrences, but also from the way all these are defined by those who examine them. The knower (scientist) and the known (nature) are not identical, but distinct. This is not a restatement of the binary separations opposed earlier in this essay, rather it is the expression of a difference as it is not possible to conceive an identity without difference. Existence comprises distinction. No known without the knower, and no knower without the known.

When the aspects/qualities of the nature (the known of the science) are classified into different/distinct divisions and some groups of scientists (the knower of science) are involved in the activity of knowing/understanding those divisions, various sub-branches of science as a branch of knowledge appear such as physics, chemistry, biology and so on. These sub-branches emanate from the investigation of nature and, though they may in time develop distinct identities, with their problems, solutions and a community of investigators, they could use some common techniques and methods such as empirical observation and experiment, objectivity, testing and so on. However, these techniques and methods, as already said, have been developed in one or other branch by some or other scientists within a socio-historical setting.

I have already made the point that science is one branch of knowledge and scientific knowledge does not consume all human knowledge, implying that there are other branches of knowledge. Precisely so. There are other branches of human knowledge arising from the distinctness and variety of the known. The reality with which we are faced and in which we are embedded is too general to form a basis for systematic knowledge. The reality, though it is a whole in this sense, could be scrutinized by concentrating on one part or one aspect. This gives way to systematic knowledge. Which part or which aspect has been or is being concentrated on is something based on the specific socio-historical situation. When some people attend to nature (one part/aspect of the reality and one of the known) scientific knowledge grows up. When some are concerned with human society/human social organization, then, history and other branches of knowledge/knowledge about society come into being. Similarly, the emergence of other branches of knowledge such as mathematics, theology and art could be accounted for. One branch, once developed, cannot incorporate other prospective branches as the existence of a branch of knowledge depends upon its distinction from the others.

History is, as already stated, one branch of human knowledge whose known is the human society itself and it cannot be encapsulated by the other branches. The past as the core of the known of the historical knowledge and consisting of particular events in space and time, which are no longer happening, cannot be comprehended by mathematical knowledge which apprehends the objects that do not have special location in space and time (what is called abstract objects). Similarly, theology cannot account for history since it is about a single and infinite object, but history is about finite and plural objects. If the known of science and history, as mentioned, are different, then, science cannot incorporate history.[46] Therefore, history forms a distinct branch of knowledge. The next chapter is devoted to the analysis of that branch.

2

History or did Napoleon Win at Waterloo?

It is a commonplace view that there may be distinguished two different senses of history. On the one hand, history stands for the totality of events, human actions and interactions, all that has happened in the past; and on the other, it refers to the knowledge or account of what has happened in the past, a particular mode of research into such occurrences of the past, as it was in the original Greek meaning of the word, 'inquiry'.[1] In the first sense, history is what man has done, and in the second, it is what man knows about, and what man makes out of, what he has done. The result of history in the first sense is, it is held, the everyday life of human beings with all its compositions, while in the second sense, it usually takes the form of history. It is in the second sense that history is said to be a branch of knowledge and, perhaps, it could simply be described as what 'historians' do, or history is, to use the phraseology of the present study, a combination of the 'historian' and the 'historical'. Saying this is not, however, saying much and the statement needs to be qualified.

Before analysing the peculiarities of history, it must be stated that what I have said earlier on knowledge in general, that knowledge is attained in and through a social (and historical) environment or, to use Kuhn's term, within a paradigm, that thinking mind (if there is such a thing) is not an empty bowl, but has its own tools, and does not operate in a vacuum, and that knowing/knowledge is presupposition-based, all go for history as a branch of knowledge.

Indeed the concept of paradigm as the constellation of assumptions, beliefs and values shared by a community and the view that our knowledge of 'new' phenomena and our theories and observations on objects, that is, the known, are filtered through that constellation, and that our knowledge is obtained via a social community are not novelties to some students of history and other subjects of human phenomena. Collingwood, for instance, insisted that the thought of any period was organized according to 'constellations of absolute presuppositions'. He also made it clear that thought was a 'corporate possession' and knowledge was socio-historically based.[2] Nietzsche emphasized the determining power of what one already has (learned) with respect to what one is to learn: 'Ultimately man find[s] in things nothing but what he himself has imported into them; the finding is called science, the importing – art, religion, love, pride.' Man 'always discovered in things only *that which he* had put *into them*.... Ultimately, no one can extract from things, books included, more than he already knows. What one has no access to through experience one has no ear for.'[3] According to Mannheim, 'all knowledge is relational, and can only be formulated with reference to the position of the observer'.[4] Donelan expresses the community aspects of knowledge and theorizing: 'All theorizing, above all, all true critical theorizing, all reasoning about the world, requires debates with others. Debate can only go on within a group that agrees roughly on the boundaries of what they are debating.' For him, knowledge is bound by the particular culture to which the theorist happens to belong.[5]

The historian is, just as anybody else, bound by his social environment, the point from which he views the events. Hegel makes it certain: 'Even the ordinary, run of-the-mill historian who believes and professes that his attitude is entirely receptive, that he is dedicated to the facts is by no means passive in his thinking; he brings his categories with him, and they influence his vision of the data he has before him.'[6] Collingwood stresses the same point, perhaps more strongly as the data/object of history is not, unlike natural science, even relatively, independent of the historian himself. He is 'a part of the process he is studying, has his own place in that process, and can see it only from the point of view which at this present moment he occupies within it'. The matter, however, goes on Collingwood, is even graver than this, for history is mainly concerned with the past and the 'historian's only possible knowledge of the past is mediate or inferential or indirect, never empirical'.[7]

Collingwood's view that the (historical) knowledge (of the past) is mediate is modally right. However, this can also be argued for other branches. There is no direct knowledge in the sense of a one-to-one correspondence between the knower (whoever he is) and the known (whatever it is). All knowledge is in principle mediate and reported. Therefore, the case for knowledge being mediate, reported and indirect can, *pace*

Hegel and Collingwood, be equally made for other branches of knowledge, even for science as shown in the previous chapter. Yet, this sentence is not intended to deny the difference between history and, say, science.

In science (nature), a fair (or relative) degree of directness in knowledge can be attainable. When it comes to society/history (human world) the degree of directness is much less for the historian is an active participant in the process he is examining (which is not the case for the scientist) and the bulk of what the historian is studying, that is, the past, is bygone. In other words, the historian and his object do not share the same moment in time. Unlike science, both the knower and the known of history are time-and-space bound. This does not mean that natural phenomena do not take place in time. What I mean is that nature, as we know it, remains as it is over long periods of time, perhaps as far back as we know. Gravity has, so it is held, been present since the formation of galaxies. This is not the case for society. Human phenomena (the known of history) are very often subject to change in time. One could, therefore, go further than Collingwood and say that historical knowledge is double mediate.

Human world and temporality

Two distinct characteristics of history have thus been suggested: it is about human world and that world is temporal, something which belongs to temporality. Emphasizing that history deserves to be an independent branch of knowledge, one which aims at 'subtle explanation of the how and why of events', Ibn Khaldun, the fourteenth-century Arab philosopher and historian, stated that history has 'its own peculiar object – that is human civilization and social organization'. History is 'information about human social organization, which itself is identical with world civilization'.[8] Ibn Khaldun does not only state that history is about human actions, but also specifies what kinds of human actions fall within the subject of history – 'human social organization' – which is equal to 'human civilization', that is to say, social actions and interactions of human beings. Another striking point in Ibn Khaldun's view is that history is not confined to a mere past. History aims at the 'explanation of the causes and origins of the existing things', implying that the starting point for historical inquiry is not past, but present.

Ibn Khaldun, who was the pioneer of the historical inquiry the present essay is set to pursue, of course, was not the only one arguing for the view that history takes human beings in social life as its subject matter.

Most of the later philosophers and historians share the same view. To give some examples, Vico, who is credited as being one of the founders of the 'rational, scientific' study of history in the modern period, said: 'Our science comes to be at once a history of ideas, the customs, and the deeds of mankind.' For Kant, history was composed of the manifestations of human will, which were determined by universal natural laws. Hegel puts the distinction between nature and history very succinctly: 'We must first of all note that the object we have before us, i.e. world history, belongs to the realm of the spirit. The world as a whole comprehends both spiritual and physical nature ... The spirit and the course of its development are the true substance of history. We do not have to consider nature here as a rational system in its own right but only in relation to the spirit.' Toynbee excluded the anthropological (meaning those societies which have no survivals coming to us) and the biological from the historical. Finally, Collingwood repeatedly made it clear that only human processes (not nature and its processes) could be the subject matter of history.[9]

History is attributed to human beings, as opposed to nature and its processes including the biological ones because, it is held, only human beings have the will and ability to reflect upon themselves, only the acts of human beings are purposive, and above all, only human beings can record their experiences. No need to say that some include natural phenomena in the scope of history. Herder, for example, begins his analysis with an account of the whole cosmos. Aron, too, takes nature as a subject matter of history in making the distinction between 'the narrow sense' of history which 'is the science of the human past' and 'the wider sense' which 'studies the development of the earth, of the heavens and of species, as well as of civilization'.[10] Those who include non-human processes within history, like Herder and Aron, however, usually make the point that nature and other non-human phenomena are counted as long as they affect human beings. Not for their own sake, but for man's sake.

So it is clear, history is about human phenomena. Following this, the question is what kinds of human phenomena could be viewed as the object of history. Do all the actions and processes of human beings constitute the subject matter of history? For Ibn Khaldun, it is those that are related to 'social organization'. Collingwood agrees with him. Although he says that only those actions, which are purposive, re-enactable, and self-conscious make history, it is said, these actions of the purposive, the re-enactable, and the self-conscious are also social. So far as man's conduct is determined by his 'animal nature, his impulses and appetites' such as eating, sleeping, and making love, it is non-historical. The historian is interested in 'the social customs which [men] create by their thought'.[11] Whatever a man in, say, the Scottish Highlands does on his own, his actions do not fall to the attention of the historian, unless they get socialized, and become a common way of behaviour in a certain community. Eating is not a part of history; but when, for example, the way that a Chinese man eats his meal, whether by hand or by chopsticks prevails, and becomes part of the table manners in Chinese society, this does become part of history. The question of how a distinction between social and non-social actions of human beings can be made is a question which one is at a loss to answer. Considering the social nature of man, as we know him, it is not hard to see how trivial the distinction is. The only possible answer, in my view, is that the distinction itself is social.

History is thus, as a branch of knowledge, a combination of the historian and human social actions and interactions. Just as it is possible to define various sub-branches of science as a branch of knowledge, there may be identified sub-branches of history according to various, identifiable aspects/qualities of the known of history – human social phenomena, such as the political, social, economic, intellectual, cultural, and so on. Today we have divisions of history, each corresponding to a particular aspect or quality of the human social phenomena, such as social history, political history, economic history, cultural history and so on. What we know as social sciences or disciplines are but intellectual activities concentrating on a particular aspect or quality of the human social phenomena. Even if it is assumed that these social disciplines are researches into different aspects of human society, the separation of these disciplines from history is impossible, as the saying goes; to understand what something is you need to know how it has come to be what it is. Braudel is right in saying that 'all the human sciences, including history, are contaminated by each other. They all speak, or are capable of speaking the same language.'[12] One student goes further and argues that there is only one discipline of the study of social phenomena, namely, 'historical social science'.[13]

It is therefore a vain attempt to separate history proper and other social disciplines. History as a general branch of knowledge about human social phenomena provides those sub-branches with, not only the background materials, but also methods and techniques and information about the nature of their particular knowns. History is closely associated with the so-called social sciences or humanities. John Seeley once said: 'History without political science has no fruit; Political science without history has no root.'[14] Similarly, one could rightly state that to study history means to study politics and to study international relations is to study history. Being well aware of this inseparability, the scholars of the past saw themselves as the students of humanities. The formation of the so-called separate social disciplines could be taken as a result of the departmentalization or fragmentation of knowledge about human society in the late nineteenth century and early twentieth century. Nothing is wrong with departmentalization: it could enhance our knowledge. What is wrong is the denial of being historical. This is where we come to the second characteristic of history, that is to say, the known of history (human social phenomena), which is inevitably 'historical', extending from the past via present to future.

Past–present–future

The last sentence is likely to raise many eyebrows. It has been a commonplace view, at least, in the literature of what may be called 'history proper' as it is taught in many university history departments, that history is concerned with past human actions, what man has done, not what man is doing or what he will do. It is about 'what actually happened' not what is happening. Not the whole of the past, however, lies within the range of history. Only those human actions of the past which are communicable to historians and those which have a link, a sort of communication system, with the present can properly be treated as the known of history.[15]

The view that only that part of the past which is knowable to us and communicable to the present constitutes history implies that history cannot be confined to a mere past. The starting point for the past is indeed the present as Ibn Khaldun stated. Historians study the past because it has an association with the present. It is not easy to separate the past from the present. Even those who strictly confine their study to the past acknowledge this, as history is said to terminate in the present. Some clearly express that history is about the past and the present, about what man has done and what he is doing as well. Von Ranke himself admitted the continuity between 'what actually happened' and 'what is actually happening'.[16]

If history comprises the past and present, then, why not bring the future as well? Just as it is not possible to separate the past from the present, it is not easy to divorce the present from the future. History is therefore composed of the past, the present, and the future. Human social phenomena have a past, a present, and a future. That history as both a process and a branch of knowledge extends from the past via the present to the future has indeed been acknowledged by some thinkers. While Braudel considers history as a collection of the special 'tasks and perspectives from the past, the present and the future', Burke declares that the state (a historical human phenomenon) is 'a partnership not only between those who are living, but between those who are living, those who are dead and those who are to be born'. Nietzsche is more explicit: 'The present and the past upon Earth – alas! my friends, this is *my heaviest* burden! And I could not live, were I not a seer of that which is to come.'[17] Such are the views of some thinkers that history can be confined neither to the past nor to the past and present.

History as known, history proper, is concerned with the past. This is certainly what most 'proper' historians think when they practise their crafts. That could be taken as a compromise of the community of historians, which comes nearest to a paradigmatic consensus in the Kuhnian sense, despite the existence of contrary views as noted above. I have no quarrel with such a notion of consensus among most historians. My contention is that history cannot be confined to a mere past; it is concerned with the present, as many admit, and has references to, and cannot be separated from, the future, as some acknowledge. History cannot be confined to a mere past; simply because historians and societies of which they are members are 'historical' beings. History is about temporality, and temporality is composed of the past, the present, and the future. The temporality of which we are first aware is the present, and the past and the future are evoked by this present, as Graham Greene hinted and Oakeshott showed us.[18] There are no clear-cut demarcation lines between the past, the present, and the future. That history as temporality comprises the past, the present and the future seems pretty obvious in our daily expressions such as 'in history there are no general laws', 'direct observation in history is impossible', 'historically speaking', and so on. In these expressions history refers to an ever-changing and continuous process.

It can be argued that history as *res gestae* may extend from the past through the present to the future, but, as a branch of knowledge, it is confined to the past and, at most, to the present. Leaving aside the question that if something and its knowledge are coextensive, and without going into the problems of 'prediction' and 'uniqueness' in history, I would say that it is true, we do not know the future, but we do know *about* the future. Collingwood tells us that 'the only clue to what man can do is what man has done'.[19] If what man has done gives us the clue to what man can do, then, it means that it can also give some clues to what man will do, for what man will do is not totally outside what man can do. It thus suggests that the future may, in some way, be anticipated by the past and the present. In his *Hobhouse Memorial Lecture* on 'The Three Laws of Politics', delivered in 1941, Collingwood comes close to this point. 'Our relation to the future is not,' said Collingwood, 'that the future, while it is still future, is to be foreknown by us; the future can be known only when it has become present; but that it has to be made by us, by the strength of our hands and the stoutness of our hearts.'[20] The future can be known when it has become present. True, yet it does not become present by itself, but with the present and thus the past. If the future has to be made by us then it means that we know about it. Otherwise, how can we make something, which we do not know of? It is men that make history. So why should it not be men that know of it?

What I mean is that we could know about the future because we, as human beings, spend our life within a society, we all are social beings. Our acts and knowledge of the future are not independent of our acts we do now and the knowledge we have now. This does not mean that the socialization process that a man has undergone will determine his future actions and knowledge. It does, however, mean that the socialization process of a man has an influence on his future deeds and knowledge. This is what I mean when I say that history, both as *res gestae* and as a branch of knowledge, denotes the past, the present, and the future. 'History proper' may be confined to the past, but it does not encapsulate the whole of historical knowledge.

Objectivist and subjectivist conceptions of history

The two different approaches to human knowledge that I have analysed in the previous chapter, namely objectivism and subjectivism/relativism, have been argued for history, too. The objectivist understanding of history refers to the view that a commonly acceptable, universally objective knowledge of history can be achievable and only history as such forms the true history. The objectivist view of history could be attributed, in one line of argument, to those who argue that in the historical process there is a generally applicable principle or an all-determining factor, usually expressed in the concept of a universal human nature, and very often, defined on the basis of rationality as the distinction of human beings. Kant and Hegel are two well-known exponents of this line of argument. 'Whatever metaphysical theory may be formed regarding the freedom of the will,' declares Kant, 'it holds equally true that the manifestations of the will in human actions are determined, like all other external events, by universal natural laws.'

Hegel follows the same route: 'World history is nearly a manifestation of the one original reason, it is one of the particular forms in which reason reveals itself.' Marx follows Kant and Hegel in making a single-cause explanation of human history.[21] No need to add that those who think that a divine authority or omnipotent being such as God or Nature governs the historical process, too, advocate a similar objectivist position. However, as Collingwood makes clear, such an understanding of history based on a universal human nature is 'possible only to a person who mistakes the transient conditions of a certain historical age for the permanent conditions of human life'.[22] But human beings do not have unaltered characteristics and conditions. They are not equipped with fixed rationality, nor even a common rationality. The attempt to understand the historical process in this kind of universalistic notion and to build an objective knowledge out of it ignores the very 'historical' nature of the phenomena and the conceptions in hand.

The second line in the objectivist understanding of history, which disregards the former as being 'speculative'[23] or 'pre-critical'[24] and rejects a universal omnipotent factor in history but assumes a common human rationality, usually proceeds from a positivistic or scientific understanding of the historical social phenomena. This view may be identified with those who are reacting against the moralizing account of history beginning with the Enlightenment and culminating in Kant and Hegel's grandiose schemes, such as Ranke; and with those who are hypnotized by the success and prestige of natural sciences, such as Buckle and, perhaps, many other practising historians. Ranke's famous dictum, that history 'seeks only to show what actually happened [*wie es eigentlich gewesen*]',[25] echoes the positivistic procedure of ascertaining facts and finding causal connections between them, for it takes for granted that 'what actually happened' is there to be discovered by the rigorous work of a historian. Elsewhere, he suggests using the positivistic technique of verification: 'From the particular, perhaps, you can ascend with careful boldness to the general. But there is no way leading from the general theory to the perception of the particular.'[26] He did not see that induction and deduction were not distinctions of different directions.

Yet, Ranke was not a positivist in the sense it is commonly referred to. He emphasized the uniqueness of historical events, which do not allow the formulation of general laws and later conceded that there is no such thing as a single 'What actually happened'. He said that 'history will always be re-written', for it was not possible for a historian to take up a pen without 'the impulse of the present'.[27] Nonetheless, he kept his belief in the achievability of a common objective history that would come as a result of rigorous, rational and critical work and with 'the hand of God' in history.[28] One could say that many practising historians take it for granted that there exists a past out there to be ascertained with rigorous, critical, patient research and thus it is possible to have an objective and commonly acceptable view of history.[29]

Science and history

It was with the impact of positivism in general and the enormous success and prestige of natural sciences that historians were led in the nineteenth century to think of their endeavour as a science, like, say, physics – but one in its infancy, no doubt, which would mature in the course of time. Buckle, one of the most passionate advocates of the science of history, argued that if human events were subjected to similar treatment as the one applied to nature, the same results would no doubt have been obtained. History has not yet become a science because of the simple fact that historians were 'inferior in mental power' to the mathematicians, physicists and other natural scientists. In other words, if men as gifted as Galileo or Newton had devoted themselves to dealing with what went by the name of history, they could have set it to rights and built a science of it.[30] John Seeley shows the way for finding regular laws in the universe: 'If we would guide ourselves aright we must register what we observe, then we must compare our observations and generalize upon them; so we shall obtain general laws, and thus the knowledge of the past will lead us to a knowledge of the future.' He also endorses Buckle's view that the science of history is still in its infancy.[31]

Many a historian, like Buckle and Sir John Seeley, strove for the formation of a science of history. They thought that the natural and historical phenomena were of the same character, or different, at most, in terms of complexity as opposed to quality. They at least hoped that scientific method would one day be applied to human phenomena as it was to the physical world. In the end, those who were infatuated by natural sciences come to a position similar to that of universalists with their belief that 'the universe obeys only "natural laws" which make it what it is' and thus in arguing for a single-cause explanation of events.[32]

The effect of science on historians was formidable. Even those who were conscious of the differences between nature and history wanted to make a science of history, whatever it might be. In his *Inaugural Lecture* (1903), Bury said that the only way to true history lay through scientific research, and declared: history is 'simply a science, no less and no more'.[33] Collingwood, who formidably argued for history as a distinctive branch of knowledge separate from science, used scientific jargon: 'Science is finding things out, and in this sense history is a science.... [T]he word "science" means any organized body of knowledge. If that is what the word means... history is a science, nothing less.'[34] The use of scientific jargon and the sensitivity on the 'scientific' character of history have not been peculiar to the nineteenth century or early twentieth century. Ritter, for example, expressed the scientific quality of history as follows: 'It is precisely this unconditional willingness, unhindered by prejudice and bias, and the tireless, rigorous critical work on the historical sources which follows from it which makes up a very essential part of [the historians'] scientific quality.' Clubb seems surer: 'History can and should be studied as a social science.'[35] Yet, such a science of history looks no more than a vain attempt.

A host of writers, Collingwood included, made the distinction between history and science. It is held that history deals with the unique and particular, and science with the general and universal.[36] This statement, however true, cannot be maintained without qualification. The historian is, like the scientist or anybody else, interested in generalizations and concerned with them. Without generalizations the historian does not, and cannot, study or explain anything at all, for all thinking (explaining), whether in natural sciences or in history or in any other branch, involves classification and classification involves generalization. What is more, the historian has to use language and language is but generalization. The difference is not in being interested in generalities or particularities, but, it is suggested, in the fact that in natural sciences, contrary to history, credence lies in generalities; that the generalizations of history are not causally connected; that history largely uses qualitative categories unlike the natural sciences which operate on a quantitative basis; that for historians generalizations which they use to understand the particular are means, whereas for the scientists who use particular facts to arrive at generalizations, generalizations constitute an end in themselves; and that generalities in history provide only general guides for the future while in science they represent specific predictions.[37] That there are significant differences of qualitative and quantitative character, of methods and aims between history and science and that one cannot encapsulate the other and one cannot be resolved into the other seem obvious enough. 'The human factor'[38] will likely remain the crucial difference between history – whether it is defined as 'what happened' or 'temporality' of the past-present – future, and science, however it is understood, either in positivistic or post-positivistic conceptions. An objectivist understanding of history, in universalistic or positivistic/scientific terms, does not look tenable. So much for objectivism. What of subjectivism?

Subjectivist view

Subjectivism or relativism in the philosophy of history, contrary to objectivism and universalism, argues for a relativistic understanding of the historical knowledge, when driven to its logical extreme consequence, similar to Feyerabend's 'anything goes' formula. The relativistic account of history may be associated with the historicism of Meinecke or Beard's relativism. It has recently been revived in the discussions on the 'new historicism' in contemporary literary criticism and the 'new philosophy of history', both of which have been grounded in post-modernism.[39] Not all of these authors could, of course, be said to have envisaged a relativistic history of an 'anything goes' type. And yet, one can find textual reference in most of them. While Meinecke stressed the determining effect of values and culture in historical explanation, Beard, opposing the Rankean ideal of objective history, endorsed that, 'what is called "objective history" is simply history without an object'.[40] Valéry, though not a historian, expressed the relativistic position in full: 'In history, I have absolutely no respect for facts, and this will continue until someone shows me that it is impossible to replace one event by another, with no trouble at all. Except for the stories, what proof have we today that Napoleon Bonaparte did not win at Waterloo? No necessity whatever. For all the facts are perforce, entirely imagined – that is they are not hard facts.' He adds that history is always arbitrary.[41]

Historicism, usually defined as the view that everything is historical (so our knowledge of the past is attained by our historical existence in the present)[42] may give way to relativism if the 'historicalness' is understood as 'being at the moment'. The historicist argument that our knowledge of history is determined by our present values and intellectual dispositions, however true, ignores the case that the 'historicalness' of the present (knowledge, values, institutions, and so on) is not something in itself, but has bearings from the past. The continuity of the historical process defies historicism as such. Moreover, it is suggested that the historicist view cannot stand logical testing. If it is true, then it is itself historical, too.[43]

In the relativist stance of recent discussions on the 'new historicism' and the 'new philosophy of history', it has been argued that there is no 'story out there' to get straight, it is always the historian's construction;

that historical narratives are verbal fictions the contents of which are as much invented as found; that there are neither past nor evidence given to us.[44] Others who are engaged in the discussions, however, challenged these relativistically loaded views and the so-called reduction of history to literature. They, evoking the Kuhnian understanding of the operation of science, have propounded a communal and consensual view of history and narratives, conceiving history as a collective enterprise with its criteria of evidence and of distinction, its rules and codes to which members are held accountable, and the standards of appraisal, coherence and veridicality.[45] Relativism of this kind, just like the old historicism, is not sustainable. In the first place, the reduction of the historical to the textual, even if this reduction is thought to be possible, is based on a misconception of the text whereby the textual is assumed to have its meaning and existence in itself. This view disregards the contextually based nature of the text. The text always takes place in a context and gets its meaning in that context, just as language has its meaning in its usage as argued and demonstrated by Wittgenstein and Collingwood.[46] The context of a text means other texts and other temporalities. Secondly, added to their misunderstanding of the textual, they seem to have a misunderstanding of the historical as well, in the manner I have already said of historicism in extreme. To say it again, the 'historicalness' does not mean being at a moment, but taking place in and going through the historical process.

Socio-historical conception of history

If objectivism and subjectivism are not tenable in history as they are not in science, then, the question is to what extent a Kuhnian framework of paradigm, a communal and consensual view, is relevant to historical knowledge. A number of historians announced the applicability of Kuhn's paradigm framework to history and its sub-disciplines, although the concept was basically put forward to account for the formation of scientific knowledge.[47] In practice, we see that Kuhn's terms have been employed explicitly by historians and the students of other social disciplines. No doubt, Kuhn's view that scientific knowledge is based on a paradigm framework which was defined by a group of scientists or a scientific community through group discussions, compromises and persuasion, and which defines and determines the methods, aims and major problems of a field (science), is most relevant to historians and their field.

The conception, that knowledge is consensual or socially based, has been familiar to historians since long before Kuhn, as I have already shown in Collingwood and Oakeshott's words. It, of course, goes further back. Herder had a hint of it and the historicists (like Meinecke) recognized it. Burckhardt stated, evoking Kuhnian paradigm choice: 'Nothing wholly unconditioned has ever existed, and nothing that was solely determinant. At the same time, one element predominates in one aspect of life, another in another. It is all a question of relative importance, of the dominant at any particular time.'[48] Collingwood did not only emphasize the socially based nature of historical knowledge and the existence of a 'constellation of absolute presuppositions', but also anticipated Kuhn's view that a new paradigm does not add just new data/knowledge, but rather, it redefines the fundamentals of the field.[49] Spengler, Toynbee and Barraclough were well aware of the fact that their analysis was conditioned by particular societies.[50] Unlike scientists, historians, at least some of them, have always been conscious of what Kuhn specified regarding the nature of (scientific) knowledge in his concept of paradigm. The question is whether the historians do operate within a paradigmatic framework or in line with Kuhn's propositions regarding the community of scientists.

It has been argued that in history there is no paradigmatic framework in which research is conducted as argued for science by Kuhn.[51] It is true that historical research does not take place in a Kuhnian normalcy. A paradigm in science, for Kuhn, comprises the agreed principles and assumptions which define the scope and limits of the field in question: rules according to which the research is carried out, which bind the community of scientists, and which the scientists follow in their activities; methods by which the research is conducted; concepts which the community use in their work; theories that are aimed to test; problems and issues into which the scientists inquire; questions which are admitted to be legitimate to ask; and goals which are commonly pursued. All these are not observable in the case of history. Historians ask different questions and pursue different goals. Unlike the tightly organized structure of the scientific community, historians are largely diversified. Nonetheless, in history, at least in history proper, there are common features. Acknowledging this, Kuhn himself calls history and other social disciplines 'proto-sciences' which generate and test – however imperfectly – 'testable propositions'.[52]

In history, despite the ongoing debates and disputes over the nature of the field, or its facts, goals, and so on, there are some points over which consensus has been achieved, at least among practising historians, such as the past being taken to be what history is about (contrary to the thesis argued here); archives constituting the major source of evidence and data; only the social or socially implicated human actions being the subject matter of history, and so forth. Certain events have been admitted to be facts, the true facts that constitute history. Therefore, *pace* Valéry, we know that Napoleon Bonaparte did not win at Waterloo. Not because it has been proved conclusively in the way the existence of gravity is proven, but just because there is nothing, as we know it, admitted against it. History is, it is very often said, what historians do. Almost all historians say that Napoleon Bonaparte was not the winner at Waterloo, assuming that the terms 'winner' and 'loser' are understood by all. Yet, there are so many issues on which historians disagree, such as the origins of the Second World War or the emergence of the 'modern international system'.

According to Kuhn, the professional community of scientists does not usually tolerate those who challenge the prevalent paradigm during the normal science period. The observation is, I think, applicable to professional historians too. One needs only to remember the reactions when Taylor's much-discussed book about the Second World War came out.[53] It is a clear indication of how resistant the professional community of historians could be, even to one who, by any standard, is a professional historian. The professional community of historians, like scientists, regularly applies intersubjective standards to the scholarship of its members. For example, Toynbee's work on Rome, *Hannibal's Legacy*, received less criticism than his *A Study of History*, especially its last volumes, as the latter contains less technical history than the former.[54] Criticisms of the last four volumes of his *A Study of History* were sharper as they contained less of technical history compared to the first six volumes. Similarly, Wells' *The Outline of History*[55] is an example of how exclusive the professional community can be in relation to non-professionals. Kuhn's paradigmatic view of science and his propositions regarding the attitudes of professional communities do have bearings for history, even though there is no paradigmatic consensus covering the whole field and no tightly organized professional community. The significance of his work for history lies in the fact that he has shown that even the scientific knowledge is consensual and thus social.

Conclusion

To recapitulate what has been said about theory in the previous chapter and history in the present chapter, I have argued that all human knowledge is social and historical for it is where we begin and end. There is no Archimedean point outside the human world on which we can rely and of which we may know once and for all. Even those kinds of knowledge that are described as intuitive or revelatory are social in the sense that in order for them to become human knowledge they have to be communicable to us; in other words, they must be expressed in a language which is a social property. Man's sociality and language make his knowledge social.

Objectivism and subjectivism/relativism can be maintained neither in science nor in history. Whereas in science the knower is historical and social, in history both the knower and the known are historical and social. This could be one reason why in history a Kuhnian normalcy is hard to achieve, but a Kuhnian normalcy, even in science, as he has shown, is not outside what is called the human world and does not equate with objectivism. Historicalness defies objectivism as it shows that nothing is universal in the sense of being independent of time and space, and socialness defies subjectivism as it shows that not everything goes. Against this binary opposition of objectivism and subjectivism, I would like to suggest 'historicism' to be an adequate account of human knowledge as I have already defined and shown.

By 'historicism', I mean an understanding of man in society who acquires his existence, knowledge and abilities within the historical process in continuity, which is nothing but men's life in socially organized collectivities in the world as we know it. Historicism as such is different from its sense made popular by Popper. History and 'historicism' are then categories that express what may be called a 'greater relativity'. Objectivism and relativism could be viewed as two reifications of that greater relativity. Historicism, in this sense, does not lead to any 'laws', fixed and applicable to the whole historical process. Yet, it does recognize patterns in history: patterns, because human knowledge and thinking depend upon generalization and thus patterning and these cognitive processes of men are not independent of his sociality, but rather moulded by it. We do not know the future shape of our knowledge and society. However, we do know what our present and past knowledge are. The socialization process, which is an important determinant of our knowledge and life, influence our future knowledge and life, thus giving way to patterns.

Knowledge and life, theory and history are not separable. Paraphrasing Sir John Seeley, theory without history has no foundation and history without theory has no presentation. The practice of 'doing' theory is necessarily historical. Nevertheless, the rejection of such binary oppositions in which one element is taken to be essential and the other marginal does not mean that there are no distinctions.

Distinctions remain. The existence or identity of a thing depends upon its distinction from other things. The constitution of our knowledge too relies upon distinctions. Without distinction no identity, without 'me' no 'others', without 'mind' no 'body', without 'theory' no 'facts', without 'object' no 'subject', and so on.

To theorize or to explain something means to historicize it, to historicize in the sense of locating and placing it within, and together with, its locality or environment. In the human world, a particular known can only be known together with the general known which has a larger spatio-temporality. The general known with a larger spatio-temporality may better be comprehended through an outlook of world history or universal history that will be examined in the next chapter.

3
Universal History or the World as we Know It

A convenient way to define universal history or world history may be to look at the unit of analysis. If the nation or nation-state is taken as the unit of analysis then the historical study could be called national history. When the world or entire globe is the unit of historical analysis, or a unit in history proper, the endeavour could be named universal history.[1] This is but one way of approaching the concept of universal history. The statement is, however, by no means self-explanatory and it needs to be elaborated on. First, what is meant by the 'world' or 'entire globe' should be defined. Second, it is presupposed that a historical analysis of the world is indeed possible.

Spatio-temporal limits

What is meant by the 'world' is naturally limited by what is known of it. The world then, spatially, comprises the known geographical limits. What is now called the Americas, for example, was not the part of the world for men living in Europe before the fifteenth century. Geographical area incontestably sets limits to what societies or social groupings of men will consider to constitute the world. Following the same example, there were no such communities as the Mexicans or Aztecs for the historians of, say, the thirteenth century. Even in the sixteenth century, Butterfield tells us, the four world-empires system, according to which the world was composed of the Assyrian, the Persian, the Greek and the Roman empires, was still employed in the schools of history.[2] No mention of China, let alone the Americas. It was only in the eighteenth century that what is called the 'Semitic' people were introduced into world history.[3] Spatially, the world or the entire globe is no more than the known area by particular societies in question.

Of course, this involves an inherent ethnocentrism or localism which is usually exclusive.

Temporally, the world could be defined as the known historical process or time-span by a particular society or man at a particular point of that process. It goes as far back as our knowledge of men who lived before takes us. It is similarly confined by our existing knowledge. For instance, it has been said that until the 1930s, the temporality of the world in terms of human civilization went back to 3100 BC, as it was assumed to have first been embodied in the civilization of Egypt. The temporality of human civilization was later traced further back with the Sumerians.[4] The origins of man, the emergence of human collectivities and the development of 'civilized' societies have always been a contested issue and given way to various conceptualizations of temporalities for mankind.

Similarly, the most common division of the temporality of 'world history' has been the threefold division of 'ancient, medieval and modern' periods. As Butterfield informs us it was first formulated by Christopher Cellarius around 1685 and it is still utilized. However, it did not go without being challenged. Barraclough considered it an 'old parochial division' and a 'meaningless time-scheme which makes nonsense of the past'. It is 'simply the idea of a mediocre German scholar three centuries ago'. For Toynbee, 'the conventional formula "Ancient + Medieval + Modern" is not only inadequate but misleading'. Spengler could not have been more in agreement with him: the current West European scheme of history, divided into 'Ancient–Medieval–Modern' periods, is 'an incredibly jejune and *meaningless* scheme'. It is a source of 'Ptolemaic system of history' that concentrates chiefly on the Western history.[5] We now know that this threefold division reflects the experience of a particular locality, that is, Europe. The world is then limited in its temporal extensions just as in its spatial extensions. It consists of limited spatio-temporalities. If it is limited, confined and particular, how can a universal history of the world, or indeed a world history, be meaningfully envisaged? This is where we come to the basic presuppositions of the initial statement.

The basic difficulty with the universal history stems from the fact that the object of it (the universe) does not seem to be readily definable compared to the 'societal' or 'national' history whose object has a more readily definable and identifiable character. The possibility of a universal history or a historical analysis of 'the world as a whole' depends upon how we understand 'the world as a whole'. To recall our formula once again, the universal history can be defined as the combination of the universe, or the world as a whole, and the historian. That in turn assumes that the universe/the world as a whole is susceptible to human knowledge and human beings/historians have the required equipment, that is to say, are capable, of knowing and understanding that whole of which they are a part. The view that the world as a whole can be known by a historian – in other words, the view that there are universally applicable elements in human history all over the world and a historian, though he is a particular person living in a particular locality of that world, could detect those universal elements – is, as would be guessed, the objectivist position I have defined in the present work. Objectivism thus provides one conception of universal history. Objectivists consider the world as a whole in terms of an entity. On the other hand taking the world as a whole to be a historical unit does not only mean the world constituting a whole, unified, single entity. The world as the known spatio-temporal extensions of human beings, by definition, involves different spatio-temporalities which hinder the existence and conception of a universal history as each spatio-temporality has its own historical development or process which precludes the existence of a single, whole world and the ability of the historian to conceive other spatio-temporalities. This is the relativist position defined in the present work and which, in its extreme, denies the possibility of a universal history.

The world has various spatio-temporalities, which are referred to as *'the known'* spatio-temporalities. This implies that various spatio-temporalities could be identified and known by human beings though they are within a particular spatio-temporality. In other words, there is the possibility of communication among those multiple spatio-temporal worlds. Furthermore, we know historically that there have been exchanges and interactions among different localities. It then follows that a conception of universal history could be advanced on the basis of the 'historical commonalities', not 'universal determinants', and the idea of the whole in terms of a perspective. This is the position of 'historicism' of the present work.

Two paths to universal history

One could, then, identify two paths to reach a conception of universal history. In the first, a universal conception of man, society, humanity and so forth is developed or assumed and its observances and operations are examined in history. In the second, complex and large units, which are thought to be relatively self-sufficient, are taken and their 'essential' features are integrated. Christian authors, Kant, Hegel and Marx follow the first path; Spengler, Toynbee, Hodgson and McNeill follow the second path. It can be said that those who follow the second path too, just as everyone else, have a conception of man, society and humanity. Yet, for them, it is not the main determinant of analysis. While the first path emphasizes unity, the second one puts emphasis on plurality. Upon these considerations, in the following pages, I shall outline the two conceptions of universal history.

Objectivist path

Of all attempts at a universal history before the 'modern' period, the most well-known are those by Christian authors, especially Saint Augustine's *City of God*.[6] The Christian account of universal history was not, of course, the first one. Collingwood traces it back to the Stoics who contemplated the idea of the whole world as a single historical unit.[7] This can hardly be taken as the original starting point. The idea of a universal history may be traced back to the beginning of history. The sense of history is something concomitant with society. *Pace* Berdyaev, who denied the sense of history to the Greeks and Hindus for having no conception of goal and purpose in the universe, one could assert that the Greek society, just as all societies, had a sense of the historical. One can even go back to the Old Testament and the Hebrews.[8]

We could find not only the traces of the sense of the historical, but also that of a universal history in the Old Testament. By connecting the historical roots of the Hebrews with the origins of mankind in the book of Genesis, the Old Testament encouraged reflection on the history of all peoples. As Coll pointed out, until the middle of the eighteenth century, almost every universal history produced by the Europeans began with Genesis.[9] The idea of Genesis was not peculiar to the Hebrews. Almost all societies have had some kind of creation epic. Furthermore, the historicalness cannot be confined to having a definite goal or purpose. Yet, the idea of universal history, in its first conception, comprises an end, or an ultimate purpose to which everything is subordinated and towards which the historical process is inevitably moving.

Judeo-Christian conception

The idea of history with an ultimate end emerges with the Hebrews and it provides them with distinction among the Ancient peoples. In contrast to the cyclic view of history held by the Greeks and Hindus, the Hebrews' conception, as it is put by Butterfield, 'provided a framework within which an idea of progress could develop' later in the course of Western civilization.[10] The idea of progress is a natural corollary of the view that history has an ultimate goal. That it was the Jews who conceived of the idea of universal history with an ultimate goal was expressed unequivocally by Ranke. He wrote of ancient Judaism: 'The steps by which this religion, when it had once made itself independent, obtained the supremacy over all other forms of religious worship, and became one of the fundamental bases both of Islam and of the Christian world, form one of the most important elements in universal history.'[11]

Ranke was right. Judaism provided an important element of universal history. It was monotheism. First formulated by Zarathustra, monotheism was put into a doctrine by the Hebrews. With the conception of one single God, who was omnipotent and himself independent of socio-historical process, but involved in it, monotheism laid the ground for a universal history. God was conceived as the ultimate universality and he was a universal determinant in history. Since all peoples were created by God, it was easy to advance universal elements in the course of human social process or history. God was in history. So, the idea that there is a judgement of God in the very process of history is older than Christianity. It was propounded by the ancient Hebrew prophets. As Ranke said, Judaism formed the basis of Christianity (and thus Islam). In that case, what is the distinctiveness of Christianity?

The Christian conception of universal history elaborated upon Judaism. First and foremost, Christianity is, so it is held, a historical religion presenting historical events and interpretations.[12] It is true Christianity is a historical religion. The question of the Incarnation, the Crucifixion, and the Resurrection may transcend the established limits of the history proper, but they imply that Christianity has rooted its assertions in that ordinary realm of history with which the technical historian is concerned. This is, as seen, a continuation of the historicalness, found in Judaism. The distinctiveness of the Christian conception comes from its being more universal compared to Judaism. There are no 'chosen people' in Christianity, but, rather, equality of all men in the sight of the God. The Christian conception of man thus transcends the boundaries of the local or particular social groupings. Perhaps, a factor behind this was the Stoic conception of man. Unlike Judaism, Christianity was, Berdyaev observed, the meeting ground of the Eastern and Western forces. It therefore supplied 'the postulate of universal history ... Establishing itself upon the ground of a united East and West, Christianity offered the postulate of a united mankind and a providence manifesting itself in historical destinies.'[13] Despite all this emphasis upon unity and equality of mankind, a schism remained in the Christian conceptualizations since non-believers or pagans were excluded from history and treated as being unhistorical.

The Christian idea of universal history, then, depends upon an omnipotent, infinite God – the only, one, single, true God. He constitutes the basis of all universality. History is directed by divine providence. Augustine's four concentric rings of human society, namely the *domus* (household), the *civitas* (city or state), the *orbis terrae* (the whole Earth and the whole human society) and the *mundus* (the universe), as pointed out by Sir Ernest Barker, depict a unified world by God.[14] God does not only supply universal elements, but, it has been argued, he also takes care of the observance of them. The Christian view of history is, writes Dawson, 'a belief in intervention by God in the life of mankind by direct action at certain definite points in time and place'.[15] It has a general conception of human nature applicable to all men and women, though it is a wicked one. Despite the seeming disorder and clashes, there is an order in the universe, because, declares Dawson, evoking Kant, 'the disorder and confusion of history are only apparent ... God orders all events in His Providence in a universal harmony which the created mind cannot grasp'.[16]

In the Christian conception, history has a beginning and an end, a creation and a day of judgement. The particular events of history can become meaningful only with respect to the ultimate goal, the achievement of which necessarily brings the historical process to a close. The end of humanity or man is 'to come to that kingdom of which there is no end'.[17] It is this final end of man that makes all particular events potentially intelligible. To sum up, the Christian conception of universal history was based on an infinite, supra-natural and supra-social being. Following this, it assumes a universality of human nature and an ultimate goal of, and a final end to, history – which is outside history. It was capable of conceiving a beginning and an end as it envisaged an omnipotent being who could stand outside history and contemplate it. It was, therefore, basically an ahistorical conception. Yet, its principal features have been preserved in what may be called 'modern' or 'secularized' conceptions of universal history.

Modern conception

The objectivist tradition set by monotheism continued to prevail in the modern conceptions of universal history which were chiefly based on, or incorporated the main features of, the developments that took place in the study of nature (what is now called 'natural sciences') from the seventeenth century onwards. The modern conceptions, like those by eighteenth century writers such as Turgot and Condorcet, the Enlightenment authors for whom the idea of universal history was particularly congenial, and later, the accounts of Kant, Hegel, Comte and Marx, modified the universal elements of the monotheist tradition. The major source of universality was transferred from God to Reason, Rationality of human being, Nature, or the inevitability of progress. It was this universal determinant that made the unity of the world and its perception of being as a whole. While the seventeenth- and eighteenth-century naturalists saw the human history proceeding under the general laws of nature, the nineteenth century authors, especially Hegel, made a clear distinction between history and nature. For Hegel, history proceeded not under the laws of nature but of Reason or Spirit. However, the source or content of universalism was conceived by them transcending the spatio-temporality in which they were living. It was still a universalism and what is more, in the account of all those authors, universalism required or led to an idea of progress, as 'all forms of universalism possess a concept of progress', rightly put by Linklater.[18]

Progress from Kant to Hegel

They all shared a fundamental belief in 'progress, development or evolution'.[19] Bury, in his still classic *History of the Idea of Progress*, defined it as the belief that 'civilization has moved, is moving, and will move in a desirable direction'. He identified it as a strictly modern product that emerged in the seventeenth century and reached its full expression in the eighteenth century, as expressed in the works of Turgot, Condorcet, Comte, Spencer and Tylor.[20] The idea of progress, *pace* Bury, cannot be taken to be a strictly modern product. Dawson, I think rightly, argued that the idea of progress was essentially a secularized version of the traditional Christian view.[21] Yet the modern conception of it may be (or is) different than the one in the monotheist tradition. In the latter, progress is governed solely by God and directed towards union with Him; in the former, it is governed by human reason, or Nature and directed towards the final stage of human happiness or the complete unfolding of man's capacities. Nevertheless, in both of them progress was towards the ideal, the achievement of which would, in a sense, bring the history to an end. Where else could one go from the ideal? The universal element/determinant and the ideal thus converge into each other outside history and the whole historical process is subordinated to that universal ideal.

To exemplify these propositions, let me take the two most ardent proponents of objectivist universal history in the modern age. In the very beginning of his seminal essay on universal history, Kant unequivocally tells us that 'the manifestations of the will in human actions are determined, like all other external events by universal natural laws', and speaks of 'a universal purpose of nature' and 'a determinate plan of nature' despite the existence of paradoxical movements and planlessness in human things.[22] The universal history of the mankind is described in his eighth proposition as follows:

> The history of the human race, viewed as a whole, may be regarded as the realization of a hidden plan of nature to bring about a political constitution, internally, and, for this purpose, also externally perfect, as the only state in which all capacities implanted by her in mankind can be fully developed.[23]

For Kant, a universal history of the world can be worked out according to the plan of nature (his ninth proposition) and, yet, he acknowledges that such 'a universal history is ... to a certain extent of an *a priori* character'.[24] Man is endowed with reason and freedom and they can only be completely developed and realized in 'the species not in the individual' (the second and third propositions) and with 'the establishment of a civil society, universally administering right according to law' (the fifth proposition). The 'unsocial sociability of men' and their states can be transcended in the final resort by the hidden plan of nature to which they are, in fact, a means (the fourth and seventh propositions).[25]

As seen, in Kant's analysis, nature is conceived to be just like the omnipotent God of, say, Christianity. It has its universal laws which govern the human actions and also a plan for man, but hidden from him. The will and reason are endowments of man by nature and conceived to be independent of human actions and their everyday life. These are therefore two universal abilities independent of the individual human beings. There are not *wills* and *reasons* of human beings, but there is *the will*, and there is *the reason*, in human beings. It is, to Kant, this triad of the nature, the reason, and the will that makes a universal history and its account possible. No need to add that it is basically an ahistorical account of history. The triad is outside the spatio-temporal realm by his self-admission in qualifying his idea of a universal history to be of an *a priori* character.

Hegel incorporated almost all propositions and ideas of Kant in his conception of universal history, with the difference perhaps, that he made a clear distinction between history and nature. 'World history as a whole,' declared Hegel, 'is the expression of the idea in space.'[26] Though Hegel made an acute distinction between history and nature, which was only vaguely stated by Kant, he described history/world history in similar terms. For Kant, it was the manifestation of the Nature, to Hegel, of the Spirit. Hegel wrote:

> Reason governs the world, and world history is therefore a rational process. Reason is substance and infinite power; it is itself the infinite material of all natural and spiritual life, and the infinite form which activates this material content... World history is merely a manifestation of the one original reason, it is one of the particular forms in which reason reveals itself... The history of the world is a rational process; the rational and necessary evolution of the world spirit... Thus, spirit is the substance of history. The world spirit is the absolute spirit... God reveals himself to man's thought... God is the eternal being in and for himself, and the universal in and for itself is an object for thought.[27]

The source of universality for Hegel, then, is 'Reason', or the 'Spirit' or 'God'. The discussion of these concepts and their meaning in Hegel's philosophy are beyond the scope of this book. What is relevant and important to us is that all are regarded as free from human social actions in time and place and thus as outside history. A universality similar to Kant's – in terms of reason/spirit being a substance and infinite power, an ultimate design, a general rationality, the existence of one original reason, a necessary evolution, an absolute spirit and an eternal being – does exist in Hegel's conceptualization of world history. In his emphasis on the distinction of history from nature, Hegel seems to have gone too far and pushed it beyond the historical process itself. It, then, as Kant's, turns out to be an ahistorical conception of (universal) history.

Based upon Kant and Hegel's accounts, the basic propositions and premises of the objectivist conception of a universal history may be summarized as follows:

The existence of a universal determinant

First, there is the idea that something universal, a universal determinant, operates in history. Stated in another way, a universal history can only be grounded in a universal being, whether it be a Single God, Nature, Reason, the Spirit, a general human rationality or whatever it is. This idea, the existence of something universal, has been so prevalent that many, though they do not adopt an objectivist position in some respects, have taken it for granted. The following examples show how prevalent the idea of universality has been among the scholars.

Ranke, the master of particularities and details, remarks that the principle of universal history is the principle of 'a common life of the human race which *dominates the nations without resolving itself in them*'. There is an ideal towards which all forces are looking. Ranke, concluding his essay on the *Great Powers*, spoke of the spiritual forces, which cannot be defined, but only glimpsed. Burckhardt wrote of 'heavenly bodies' shining upon all times and all peoples, of 'the same great universal tradition' perpetuating every nation. Even for Toynbee, who first conceived a universal history in terms of the comparative study of civilizations which were regarded as self-sufficient, 'history is the interaction of God and Man. History [is] a vision of God revealing himself in action to souls that were sincerely seeking him ... History acquires a spiritual meaning when man catches in history a glimpse of the operation of One True God.'[28]

A common conception of human nature

Second, as the corollary of the first premise, a common conception of human nature is advanced. It is assumed that all human beings, due to the universal element, have some common features to form what is called human nature. It could be qualified as being rational, good, evil and so on. Human nature is then understood to be a unifying element of human beings. It lies behind the essential oneness of mankind in terms of morality. Mankind, as one biological species, has a moral oneness in addition to its biological oneness. It has indeed been argued that the Stoics and the Christian authors built their doctrine of the moral oneness of humanity upon the biological oneness of mankind. It seems to me that moral oneness of mankind in Christianity comes from the idea of a single God rather than biological oneness.[29]

Nevertheless, it is assumed that as a consequence of this common human nature which has been, is, and will be, the same for everyone, humanity forms a whole despite various differences and changes in races, cultures, customs, and so forth. The conception of a common human nature constitutes one of the bases of many a historical analysis of objectivists, and even others, too. For instance, Herder wrote: 'The Nature of man remains ever the same.' He has born with passions, is born and will be born with passions. Even Butterfield, though he acknowledges that the 'eliciting of general truths or of propositions claiming universal validity is the one kind of consummation which is beyond the competence of history to achieve', agrees with the view that envisages a generality of human wickedness. Walsh too endorses the view that a common human nature does exist.[30]

The unity of history

Third, after the universal determinant and a universal/common human nature comes the idea of the unity of history, in the sense that all historical events and interactions constitute one single whole, a whole in terms of an entity. The unity of history is recognized not in particular events, but behind and above them. For Kant, the unity of history can only be seen in the whole species, and, to Hegel, the 'universal element is not to be found in the world of contingent phenomena; it is the unity behind the multitude of particulars'.[31] For Kant and Hegel, it is not difficult to contemplate a unity of all historical events, if one considers their attribution of the whole historical process to Nature or the Spirit. Ranke formulates the unity of history in terms of a hierarchy of causes, extending from 'pragmatic causality' to God.[32]

The idea of the unity of history leads to the idea that there are wider, common, spiritual forces transcending individual facts and societies and nations. In other words, the particular fact or society does not only become intelligible within the totality of facts or general society, but is also determined by the general/universal. In Kant and Hegel this is obvious. Berlin adds Marx to them.[33] Ranke's essay on the great powers is an expression of this view. In criticizing the studies on 'Universal History' in England, Ranke stressed the general connection of things and alleged that universal history should go beyond a collection of national histories.[34] Yet, he was no holist in a determinate sense as he was very keen on details. Even he argued that we know those laws and forces through studying the particularities. For Ranke, universal history should be grounded in both the general and the particular.[35]

The view that there is a unity in and behind history and that there are wider forces making it a whole cannot, of course, be confined to the objectivist account of universal history. A host of other writers also share the same view. The existence of a wider and higher element in history and the subordination of individual nations/societies to it have been expressly stated in Lord Acton's Letter sent out to the contributors of the *Cambridge Modern History* in 1898. The much-quoted passage is worth quoting again:

> By universal history I understand that which is distinct from the combined history of all countries, which is not a rope of sand, but

> a continuous development, and is not a burden on the memory, but an illumination of the soul. It moves in a succession to which the nations are subsidiary. Their story will be told, not for their own sake, but in reference and in subordination to a higher series, according to the time and the degree in which they contribute to the common fortunes of mankind.[36]

What is, however, distinctive of the objectivist position is that it conceives unity and whole as determinative of individual events, human actions and human societies. Human social actions are not regarded as taking place within a whole in mutual interaction, but as being determined by the whole.

The continuity of history

The fourth proposition of the universal history conceptualized in objectivist terms is the continuity of history. This idea of the continuity of history may imply (a) the movement of history along a certain line or in a certain direction, (b) the operation of the same principles, forces and laws throughout human history and (c) the inclusion of all times and peoples in the conception of history. All these meanings could easily be detected in the objectivist tradition. That history proceeds through and towards a certain line and direction is the natural outcome of the idea of progress and the view that there is an end in history. That there are general principles, forces and laws working throughout history is an obvious conclusion of the conception of a universal element and determinant. The inclusion of all times and peoples results from the view that there is a universal element in history and that the world constitutes a single whole.

Of those three implications of the continuity of history, the most significant one has been the idea of progress. On the way to the progression, to the manifestation or the realization of the universal spirit or goal, the means is the 'unsocial sociability of men', for Kant; the individuals or the state which is the unity of subjective will and the universal, for Hegel; and class struggle, for Marx.[37] The idea of the continuity of history cannot easily be confined to the objectivist tradition. History, by definition, implies continuity, at least in terms of the passage of time. With the state of the world and the universe as we know it, and the notion of time that we have, human social life continues. In this sense, every one has a notion of continuity. However, the objectivist makes the continuity of time and social life *in a certain form and in a certain direction.*

The ultimate end and meaning of history

A final premise of the objectivist understanding of the concept of universal history could be expressed as the persistent notion that in history there is an ultimate end which gives it (history) its meaning. Although this premise has already been implicitly expressed within the previous propositions, it needs to be further elaborated. According to this premise, there is an end of history towards which humanity as a whole is evolving and history gets meaningful by virtue of that final end. The end or goal of history is taken as a reference point for the meaning of/in history. In other words, the core of the question lies in the search for a meaning in/of history. The historical process should, so it is held, be meaningful. All this cannot be for nothing and in vain. Then, what is needed, so the argument goes, is something which could give history its meaning. The objectivist conceives the meaning of history by referring to something outside the historical process, as already mentioned, such as Nature, the Spirit, God or the final stage of human society of which one cannot speak in historical terms. The idea behind this conception is the belief that the meaning of history could not be attributed to temporary and transitory phenomena.[38] Plus, there is the view that something gets its meaning from something else. That is why, in the accounts of objectivists outlined here, we see a universal determinant.

The search for meaning in/of history is a legitimate endeavour. *Pace* many professional technical historians, who have disregarded the question of the meaning of history and brought that endeavour to a 'standstill', in the words of the lamenting historian,[39] by treating it as a metaphysical or theological issue and outside the scope of history as they understood it in strictly empiricist terms, the question 'what is it all about?' persists. The reason why the objectivist looks for something supra-historical for the meaning of history and why the technical historian disregards the problem is that they both have a misconception of the word 'meaning'. For both of them, the meaning is given. The meaning of something is not its possession, but given to it. Following this logic, one line of argument takes the already stated view that searches for meaning in history outside and beyond the history itself as it was supposed by Christian authors like Kant and Hegel. That history has its meaning through something beyond and outside it has been a relentlessly recurring theme. 'We must', proclaimed Barraclough, 'seek for history an end outside itself – as it had, for example, when it was viewed as a manifestation of the working of God's providence. That statement is not intended to imply a return to a theological view of history…, but it does mean that its study should have a constructive purpose and a criterion of judgement, outside and beyond the historical process.'[40] Yet, he does not tell us what that thing, 'outside and beyond the historical process', is. The other line of argument refutes the question on the grounds that we cannot display the value and meaning of history as a whole, 'for who could get outside of history to contemplate it?'[41]

As I have already stated, both lines of argument rely upon a misconception of meaning. True, the meaning of something is not by and in itself. But this goes for the others as well. The meaning of something, like its existence and identity, comes from its being with the others, not just from the others. Even God, supposing he does exist, becomes meaningful, at least to us, when he is historicized. Contrary to Barraclough's proclamation, it is very unlikely to have something 'outside and beyond the historical process'. Those who argue that history as a whole has no meaning because no one could get outside of it to contemplate its meaning or value, make the similar mistake. It is right, nobody can get out of history, but there is no need for that. To repeat: nothing gives anything its meaning, everything gets its meaning in interaction with others. Furthermore, as long as we could speak of history or history as a whole, and as long as the word 'meaning' makes sense to us or exists in our language, history (as a whole), *pace* Kuzminski, has a meaning.

To sum up, objectivist accounts of universal history or world history with their premises, outlined above, are but de-historicizations of the historical. The first two premises (the existence of a universal determinant and a common human nature) are neither logically nor empirically demonstrable. The last three have been built upon the first two. Consequently, human history seems to have been pushed out of the spatio-temporal realm. Universal history based on objectivism thus becomes the reification or universalization of something particular, as it was in the case of the 'jealous god' of Judaism (indeed all monotheist religions) that denies deity to all the other gods. The schism, seen in the Christian conception, thus reflects in the modern conceptions in the form of 'civilized' and 'uncivilized', 'progressed' and 'backward', 'Europe' and 'non-Europe' and so forth. Needless to say, the latter is treated as unhistorical and its historicity denied.

As Butterfield rightly stated, the study of universal history has been largely Eurocentric. In the eighteenth century, it was confined to the biblical, the classical and the European states-system. In the nineteenth century, it comprised only the continent of Europe and its relations with the world overseas.[42] This Eurocentrism (more properly, localism) has been influential upon even those who acknowledged the distinctive features of different societies and civilizations. For example, Herder thought that non-European societies were essentially static and irrelevant to the study of historical process.[43] Ranke agreed with him entirely: 'At times the conditions inherited from ancient times of one or another oriental people have been regarded as the foundation of everything. But one cannot possibly use as a starting point the peoples of eternal standstill to comprehend the inner movement of world history.'[44] The conception of universal history in objectivist understanding becomes, at the end, not only ahistorical, but also exclusive. Universalism ends up in parochialism.

I have shown that the objectivist conception of universal history/world history can hardly be taken as historical. Relativism/subjectivism, by definition, is not a proper basis for an understanding of universal history or world history. With its attribution of a self-sufficiency and omnipotency to the particular/individual, relativism hinders the world as a whole being taken to be a unit. In the remainder of this chapter, I would like to argue for a historical conception of universal history.

A historical account of universal history

It has already been said that there have been those who attempted to make a historical account of universal history. Herder and Ranke, notwithstanding their universalistic views, are the foremost. Toynbee and Spengler, despite the latter's relativistic assertions, have been considered as the twentieth century revivers of the conception of universal history or world history 'on new foundations' with their comparative study of civilizations.[45] Following Toynbee, Barraclough argued for the necessity of a conception of universal history and McNeill advanced 'organizing concepts for world history' based upon pattern of group interactions which he classified as hunting-gathering, barbarism and civilization. McNeill takes group interactions together with the geographical or natural conditions.[46] Similarly, taking group interactions and exchanges among the four civilizations of the Afro-Eurasian zone, namely Europe, the Middle East, India, and China and the Far East, Hodgson argues that they constitute 'the whole Afro-Eurasian complex' which is the only context for an adequate conception of world history.[47]

More recently, Zagorin has argued that it is not possible to attain a total conception of world history or the historical process in objectivist terms. Nevertheless this does not preclude the feasibility of focusing on large-scale subjects at a quite general level and on questions that surpass the specialist and disciplinary boundaries. It is possible to have an understanding of whole societies and civilizations and of broad areas and aspects of the past.[48] There is no need to extend this list. It is simply a indication of the possibility of a historical conception of universal history. One can even find some historical insights in objectivist propositions above, in the last three. Drawing upon all these accounts, the basic points of a historical understanding of universal history could be outlined as follows:

The known world

I have already made the point that the world for us is what we know of it. First and foremost, a historical conception of universal/world history could therefore be developed only on the basis of the known world and the process through which that world has come to be as it is known. In the beginning of this chapter, it is said that universal history takes the world as a whole which is an all-inclusive unit. The all-inclusive unit is nothing but the institutionalization of the sum of all possible units. The unit that is all-inclusive is then a known to us as all possible units can only be comprehended to be possible if we have known of them, if there is a relation of familiarity between us and what we speak of.

The known world, on the other hand, can only be known if there is sufficient flow of relations between us and others, among the existing various social groupings. The flow of relations needs to be sufficient to the extent that it allows us to become familiar with the others. It follows that it is very unlikely that we could conceive a world history which is once and for all and which is beyond what we know of the world.

The particular and the general

Second, universal history starts not from a universal, but at the point at which the historian finds himself. The point at which the historian finds himself constitutes the ground for all human knowledge. This is how Toynbee constructs his study of universal history when he begins his analysis by deconstructing Great Britain, the point at which he finds himself.[49] And it was also implied by Ranke when he boldly declared that 'the particular bears the universal within itself'.[50] If the historian departs from, or is bound to depart from, the point at which he finds himself, and the point at which he finds himself is a particular point, then, the logical question is, how could one possibly develop a 'universal', or 'general' or 'wider', history from a particular point? Ranke was quite sure of it: 'From the particular, perhaps, you can ascend with careful boldness to the general. But there is no way leading from the general theory to the perception of the particular.'[51] Provided that this is not conceived in terms of the opposition of inductivism versus deductivism, in which one part of the dichotomy is disregarded and the other is deemed to be essential, Ranke may be right, though he seems to have conceived of it as such.

To start from the particular, at a point in space and time, does not preclude the knowledge of the general and the other points. In order to start from the particular, one needs to know what and where the particular is, and to know the particular and a specific point, one needs to be aware of the general and other points. In the words of a modern-day interpreter, 'one cannot be conscious of a "here" without at the same time being aware of a "there" '.[52] It is true, 'the particular bears the universal within itself', but not in the form of the universal manifesting itself in the particular as Ranke and others have understood. The particular bears the universal within itself, just because it is perceived as 'the particular'.

Unity through diversity

Next to the question of the possibility of a universal, general conception of history is whether the world or the historical process as a whole has some universal, common, similar features or developments so as to allow us to make a general account of it. Here, we come to the third point which is the unity or wholeness of the world. A unity or wholeness of the world could be argued, I think, on the basis of empirical, historical developments in the world as we know it, not on the basis of a universality unifying the world and historical process and making it a whole entity as it was assumed in the objectivist understanding.

First and foremost there are some features or needs common to all human beings, what are called physical or biological needs. As far as our knowledge goes, every human being has to eat, drink and sleep. Life itself, death, at least for now, and the Earth, at present, are common to human beings. Second, even if we may not be as sure as Ranke in asserting that 'no people in the world has remained out of contact with the others'[53] we could see the similarities or commonalities among different social groupings emerging in the course of history. McNeill shows us that such commonalities have been existent among different civilizations, coming out of their interactions, and they have been frequent and worldwide after AD 1000.[54] Furthermore, if there are different social groupings or civilizations, and if we are able to note the differences, then it means that there are similarities and commonalities among them. It is in this sense and because of the nature of our knowledge of the historical that there is a unity in history. Ranke is right in making the startling statements that 'history is by its nature universal' and 'the union of all depends on the independence of each'.[55] The unity of all depends upon the independence of each because if the 'each' is not independent we cannot speak of the 'all'. The unity of the world in terms of a historical understanding is then, to use the clichéd statement, a unity through diversity.

Continuity and difference

The unity of the world comprises not only horizontal extensions, that is, multiple social groupings, but also vertical extensions, that is the past as far as we know it. Universal history should be universal both in time and space. It should take the past as a whole.[56] This, the idea of continuity, could be taken as the fourth point of a historical conception of universal history. The continuity is seen not along a certain line or direction and towards a certain end or purpose. It is understood with reference to the socialization process that all human beings undergo and to the accumulation of knowledge that all human beings have. It is not just the fact that time passes, but, more significantly, that throughout that process, men are in society and different social groupings are in interaction. It does not mean that there is no difference between two ages, between two periods of the human social life. Yet, as Freeman observed, 'the fact that we note the point of difference is the surest proof of essential likeness'.[57] As in the case of unity, in order to see continuity one needs to perceive different periods.

A perspective of the whole

The notions of unity and continuity are, as already made explicit, related to the idea of the whole. The idea of the whole denies the omnipotence or self-sufficiency of the individual element. The conception of universal history then considers the interrelated and relevant events or social groups. It is 'a search ... beyond the immediate facts'.[58] The idea of the whole in terms of an entity, followed to its logical consequences, eliminates the identity of elements. Without elements, no whole exists. The idea of the whole can hardly be taken in a determinate manner. A whole having a one-way determinative capacity is not human and historical; it is nothing but reification of something particular. So, only the idea of the whole in terms of perspective could be an adequate basis for a universal history. Depending upon this view, we could agree with Barraclough that the world history does not necessarily encompass the whole globe; we can perfectly and legitimately have a 'world history' of Europe alone. In other words, European history is interpreted in relation to its place in the world.[59] The European or, for example, Indian history is not considered in itself, but with others.

Universal history as such cannot be centred on a particular locality. It should be purged of all parochialism. Universal history as part of a historical understanding, as it is seen here, is neither Eurocentric nor state-centric. The civilizations of China, India and Islam are just as much part of the historical background of our times as is the civilization of the West. As Stavrianos asserted it, the basic rule that must be kept in mind is that 'no European movements or institutions [should] be treated unless non-European movements or institutions of similar magnitude and world significance also be treated'.[60]

Yet a conception of universal history is not a substitute for European or national history. As stressed by Webster, as long as the world is organized in terms of nation-states and the state has so great an influence in determining contacts among peoples, state and state relations will be part of history.[61] This observation, made in 1933, is not irrelevant in our own day. This does not mean that the world will never be able 'to escape from the historical limits of state'.[62] It is precisely a conception of the universal history as such that transcends the so-called historical limits of the state. State is but one form of social grouping, or social organization. By recognizing this, that is, being conscious of the fact that the state or any other form of organization, is just a particular form, we can recognize the existence of others. The concept of universal history may then help us to understand others.

A historical conception of universal/world history is not defined according to a universal determinant, and nothing is an island. It follows that it can only be constructed on the basis of, or through, historical commonalities. One way of examining these historical commonalities is through the comparative study of civilizations, which are large-scale complex groupings of human beings. However before we discuss civilizations we first need to consider the notion of civilization. The following chapter examines the concept of civilization.

4
Civilization or Naked Greed

The root word of 'civilization' is *civis* (citizen) or *civitas* (city) in Latin. Among other things, the term, in its Latin root, basically refers to the state of being related to, of pertaining to, of belonging to a collectivity of people, an organized collectivity, a body politic that we may call a state or commonwealth. It refers to the city-life or 'citification' or process of 'civilization' in the social life of human beings. In Latin, then, the word 'civilization' is closely associated with 'city' or 'citification'. The association of the word 'civilization' with 'city' can be found in other languages as well. The Turkish word for 'civilization', '*medeniyet*', in its Arabic root, too, has an association with 'city'. The word for city in Arabic, '*medina*', has the same root as *medeniyet*, that is, *mdn*. The other word in Arabic for civilization, '*umran*', used by Ibn Khaldun, is derived from a root that means 'to build up, to cultivate'. Ibn Khaldun uses it to designate any settlement. The association of the word 'civilization' with 'city' is by no means common to all languages. For example, the Chinese word *wen* (for civilization and culture) does not imply city or city-life.[1]

Etymologically, the word 'civilization' may be associated with city or citification. The meaning of a word can, however, only be derived from its usage. In order then to understand what is meant by the concept of 'civilization' we first need to see how it is used.

The word 'civilization' and its usages

The word 'civilization' in French and English, with its modern usage meaning 'the state or process of becoming civilized', first appeared in the eighteenth century. It was born out of the verb, 'to civilize', and the participle, 'civilized', which had long existed and were in common use in the sixteenth century. A couple of works or authors have been cited as the first literary evidence of the appearance of the word civilization in its modern sense. According to Braudel, it was first used by the French statesman and economist Turgot in his draft on universal history in 1752. The official appearance of the word in print first occurred in 1756, in a work entitled *A Treatise on Population* by Victor Riquetti, Marquis of Mirabeau.

L. Febvre and Elias agree with Braudel. For the former, the earliest printed use of the word civilization is by Boulanger in his *L'Antiquité Devoilée par ses Usages*, printed in Amsterdam in 1766. The latter, Elias, finds the first literary evidence of the evolution of the verb *civilizer* into the concept *civilization* – referring to softening of manner, urbanity and politeness – in the work of Mirabeau in the 1760s. Another author using the word about the same time is said to be Baudeau in his *Ephémérides du Citoyen* (1767). The word entered into the *Academy's Dictionary* in 1798. Contrary to this agreement upon the eighteenth century as the birth date of the word civilization, a much earlier date has been suggested. According to Wundt, it was Bodin who first used the word in its modern sense in the sixteenth century.

The *Oxford English Dictionary*, too, gives first citations of 'civilization' from the eighteenth century onwards. The word is defined in three senses. The first, and the earliest, is a technical sense in law: 'A law, act of justice, or judgment, which renders a criminal process civil; which is performed by turning an information into an inquest, or vice versa.' In this sense, it is cited from 1704 onwards. Secondly, the word means 'the action or process of civilizing or of being civilized' and it is cited from 1775. In its third sense, 'civilization' denotes, more usually, 'civilized condition or state; a developed or advanced state of human society; a particular stage or a particular type of this'. In this sense, it is first cited from 1772 onwards.[2]

Connected words, 'civility', 'civilize' and 'civilized', were used earlier than the word 'civilization'. The *Oxford English Dictionary* gives four meanings of 'civility' as 'connected with civilization, culture': (1) 'The state of being civilized; freedom from barbarity' (cited from 1549 onwards); (2) 'Polite or liberal education; training in the "humanities", good breeding; culture, refinement' (cited from 1533 onwards); (3) 'Behavior proper to the intercourse of civilized people; ordinary courtesy or politeness, as opposed to rudeness of behavior; decent respect, consideration' (cited from 1561 onwards); (4) 'Decency, seemliness' (first cited in 1612). For Elias, the concept of *civilité* received the specific stamp and function in the second quarter of the sixteenth century. It owes the specific meaning to a short treatise by Erasmus of Rotterdam,

De civilitate morum puerilium (On Civility in Children), which appeared in 1530.[3]

Six meanings of the verb 'to civilize', from which 'civilization' is derived, are listed as follows, with citation dates in parenthesis: (1) to bring out of a state of barbarism, to instruct in the arts of life and thus elevate in the scale of humanity; to enlighten, refine and polish (1601); (2) 'to make "civil" or moral; to subject to the law of civil or social propriety' (1640); (3) 'to make lawful or proper in a civil community' (1643); (4) 'to turn a criminal into a civil cause'; (5) 'to become civilized or elevated' (1868); (6) 'to conform to the requirements of civil life, to behave decently' (1605). The participle 'civilized' is defined in two senses. The first one refers to being 'made civil; in a state of civilization' and is cited from 1611 onwards. The second one is defined as 'of or pertaining to civilized men' and cited from 1654 onwards.

Civilization and culture

The term civilization is very often accompanied by the term culture, even though they are not synonymous. Sometimes, they have been used interchangeably as in 'Western civilization' and 'Western culture'. It would therefore be helpful to have a look at the word 'culture'. Compared to civilization, the word 'culture' has a longer history. Braudel says that even Cicero speaks of '*cultura mentis*'.[4] 'Culture' is derived from Latin *cultuae*, from the verb *colere*, with the meaning of tending or cultivation. In Christian authors, *cultura* has the meaning of worship. The Old French form was *couture*, later replaced by *culture*.

In English, the following usages can be noted: 'the action or practice of cultivating the soil; tillage, husbandry' (1420); 'worship; reverential homage' (1483); 'the cultivating or development (of mind, faculties, manners, etc.); improvement or refinement by education' (1510 More, 1651 Hobbes, 1752 Johnson, 1848 Macaulay); 'the training of the human body' (1628); 'the training, development and refinement of mind, tastes and manners; the condition of being thus trained and refined; the intellectual side of civilization' (1805 Wordsworth, 1837 Emerson, 1869 Arnold); 'a particular form or type of intellectual development. Also, the civilization, customs, and artistic and other achievements, of a people, especially at a certain stage of its development or history' (1867 Freeman, 1871 Tylor). Kroeber and Kluckhohn note that the word 'culture', in almost all its usages, retains 'the primary notion of cultivation or becoming cultured', derived from its Latin root. Yet, based upon their survey of anthropological, sociological, psychological and other relevant (historical, educational, and so on) literature, the authors cite 164 meanings of the word 'culture'.[5]

From this summary of the various meanings of civilization and culture what we realize, in the first instance, is that the two words have had a close association, and sometimes referred to the same thing. By the mid-nineteenth century, the two terms came to be used interchangeably in the literature of anthropology and ethnology. It became common after Gustav Klemm's (1843) use of the German word *'kultur'* to include the French term *'civilization'*, for there was no word for 'civilization' in German. Especially, with the adoption of Klemm's usage by Tylor (1871), the two words became, in a sense, inseparable in anthropology, even though most anthropologists preferred to use 'culture'. Yet, by 'culture', anthropologists mean also what may be included in 'civilization'. The earliest treatises of culture (civilization) are said to be C. Meiner's *Grundriss der Geschichte der Menschheit* (1785) and G. Klemm's *Allgemeine Cultur-Geschichte der Menschheit* (1843), though each recognized predecessors going back to Voltaire.[6] The recognition of Voltaire as the first historian of culture (or civilization) is a commonplace, even though he did not use the word civilization. It is held that he was the first scholar who examined the whole life of societies, not just the dynasties, kings and their battles.[7] If someone is to father the study of civilization (culture), it is not easy to trace who he is. Nonetheless, we could assert that it was not Voltaire, as Ibn Khaldun preceded him.

Positive and negative connotations

It was not only the anthropologist or ethnologist who included what is meant by 'culture' and 'civilization' in one word – either 'culture' or 'civilization'. Many scholars did the same so that it would not make a considerable difference if the words were replaced by each other. In fact, it could be said that both words fairly established themselves by the beginning of the nineteenth century. Guizot, in 1828, confidently proclaimed that 'civilization is a fact like any other'. It was a fact susceptible to being studied, described and narrated. It constituted a fact *par excellence*, 'the sum, the expression of the whole life of nations'. Civilization as a fact was equated with progress and development. 'The idea of progress, of development,' says Guizot, 'appears to me the fundamental idea contained in the word, *civilization*'. Civilization denotes, on the one hand, the development of society in terms of an increasing production of the social strength and happiness and also in terms of a more equitable distribution, among individuals, of the strength and happiness produced.

On the other hand, civilization means the development of the individual, of his faculties, his sentiments, and his ideas.[8]

Guizot's bold statements, confidently expressed, show that the concept of civilization has been well established. They can also be taken as the reflection of the self-confidence of a rising Europe. Guizot thinks that civilization is *ipso facto* valued. He never questions if it is something good. However, Mill, only eight years after Guizot, asks whether civilization is a good. By 'civilization', he too means 'human improvement'. According to Mill, there are two basic characters of a state of high civilization: the diffusion of property and intelligence, and the power of cooperation. Civilization is, on the whole, a good; though he speaks of some negative effects coming from civilization.[9] Mill is not as sure as Guizot. The English cautiousness? Perhaps, yes. De Gobineau (1853–55) kept the value-loaded meaning of the concept. He defined civilization as 'a state of relative stability, where the mass of men try to satisfy their wants by peaceful means, and are refined in their conduct and intelligence'.[10]

In all these treatments of civilization, what comes out is that civilization is taken to be both a process and a condition, or property, of man and society. Generally, it is ascribed a positive qualification, perhaps since its inception. The positive connotation of the concept has by no means been commonly accepted. Mill hesitated about it. By the mid-nineteenth century, it was openly questioned. Marx and Engels wrote in the *Manifesto of the Communist Party* (1848): 'There is too much civilization, too much means of subsistence, too much industry, too much commerce.'[11] Later, in *Origin of Family, Private Property and State* (1884), Engels spoke in sharper words: 'Naked greed has been the moving spirit of civilization from the first day of its existence to the present time; wealth, more wealth and wealth again ... was its sole and determining aim ... The exploitation of one class by another is the basis of civilization,' so, 'its whole development moves in a continuous contradiction'.[12] Marx and Engels' remarks could be seen as an expression of, or attention to, the likely side effects of technical progress or industrialization which has been regarded as a major component of civilization.

The qualification of civilization to show disapproval, as made by Marx and Engels, or others for that matter, may be taken, according to Braudel, as an expression of the duality between spirit and nature – a duality that has been tenaciously persistent in German thought. Culture in the German language, from Herder on, meant scientific and intellectual progress freely removed from any social context. It referred to a set of normative principles, values and ideals. By civilization, the German language simply intended the material aspect of man's existence, denoting a mass of practical, technical knowledge, a series of ways of dealing with nature. In this dichotomy, Braudel argues, one word is devalued, the other is exalted. The word 'culture' assumed the dignity of spiritual concerns, the word 'civilization' denoted the triviality of material affairs.[13] According to Elias, the French and English concept of 'civilization' can refer to political or economic, religious or technical, moral or social facts. The German concept of '*kultur*' refers essentially to intellectual, artistic and religious facts, distinguishing it from political, economic and social facts.[14] For Spengler, the civilization which is 'the inevitable destiny of the Culture' signifies the death of a culture.[15] The qualification of civilization and culture according to some sort of dichotomy is not just something peculiar to German thought. The traces of that duality may be found in the works of many scholars, from Arnold's *Culture and Anarchy* (1869) to Snow's *The Two Cultures and the Scientific Revolution* (1959). Devaluation or exaltation, the concept has usually been value laden.

Of course, there have been those who defined the concept of civilization, united with culture, in a 'technical' sense, dissociated from being value laden as much as possible. Tylor's definition, which became established in the mainstream literature of anthropology, might be regarded as an example in this direction. 'Culture or Civilization,' wrote Tylor in 1871, 'taken in its wide ethnographic sense, is that complex whole which includes knowledge, belief, art, morals, law, custom, and any other capabilities and habits acquired by man as a member of society.'[16] Tylor, as seen, took the terms 'culture' and 'civilization' as identical and, indeed, used them interchangeably throughout his work. If the arguments of the earlier chapters of this book are recalled, it is hardly feasible to expect Tylor's, or anybody else's, attempt to be successful in making civilization devoid of any value. In technicalizing the word, Tylor did not render it free from any connotation, positive or negative. What he did was to include within the concept both connotations. Tylor's definition could be viewed in line with that of Ibn Khaldun who seems to define it as what man, as a member of society, has done and has been doing. Ibn Khaldun considers the *bedouin* life (nomadic life) to be the ground for settled life and sedentary culture in which civilization grows longer.[17] The 'technical' definition, as provided by Ibn Khaldun and later Tylor, could be taken as an adequate definition in practical terms. As I have already said, it is not free from value judgement, but at least it makes the concept free from reifications as is seen in such phrases as 'civilization is progressing', 'civilization has overcome', 'civilization penetrated' and so on.

Just as there have been attempts to dissociate civilization from a value-laden content, we have witnessed the attempts to distinguish the concepts of culture and civilization. A distinction appears to have been prevalent, expressed in one of the definitions of the word 'culture' by the *Oxford English Dictionary* as 'the intellectual side of civilization'. By this distinction, attributed to the persistent duality in German thought by Braudel, culture refers to 'intellectual' advancements and achievements and civilization denotes 'material' advancements and achievements. Needless to say that such a distinction depends upon the separation of the intellectual and material, which is, as I have argued, untenable.

A second prevalent distinction is to treat one concept as a general and inclusive category and the other in terms of the subcategories of the former. This distinction is a distinction, as the saying goes, in degree not in kind. Most of the writers have taken civilization as a larger category and culture as the component of civilization. While Braudel takes civilization as 'a collection of cultural characteristics and phenomena', Elias makes the point that the concept of *kultur* delimits, whereas the concept of civilization plays down the national differences between peoples.[18] Melko straightforwardly expresses and considers civilization to be 'large and complex cultures, usually distinguished from simpler cultures by a greater control of environment', including the practice of agriculture and the domestication of animals. Civilization incorporates a multiplicity of cultures.[19] A civilization is, according to Hodgson, a compound culture, 'a relatively extensive grouping of interrelated cultures in so far as they have shared in cumulative traditions in the form of high culture, on the urban, literate level'.[20]

What we see here is that, on the one hand, civilization is a collectivity of multiple cultures; on the other hand, culture is seen as constitutive of civilization. Culture could therefore be more inclusive as well. Bagby suggests a distinction on this ground. He defines culture as 'regularities in the behavior, internal and external, of members of a society, excluding those regularities which are clearly hereditary in origin'. Civilization is, he says, 'the kind of culture found in the cities'. Cities are in turn defined as the agglomerations of dwellings many of whose inhabitants are not engaged in producing food.[21] Such attempts to distinguish the terms 'culture' and 'civilization', though they seem more tenable than the 'intellectual' versus 'material' distinction and could serve pragmatically, can hardly be maintained, for each word can safely be replaced by the other without losing the meaning of both and causing much confusion. The difficulty and artificiality of the distinction between the meanings of culture and civilization has been noted by some.[22] The distinction as such could be useful for pragmatic purposes, but no more. Besides, civilization and culture, however they may be distinguished, refer to what man himself has produced and inherited from the previous generations.

Three meanings of civilization

If I sum up so far, three distinct, but interrelated, meanings of the concept of 'civilization' can be said to exist in common usage, though their contents may vary.

(i) Civilization as a quality

The first meaning is the adjectival form. Here, civilization qualifies men and society. It refers to the state of being civilized, to the possession of good manners and self-control. When we say phrases like 'a thoroughly civilized man' or 'in a civilized country', it is in this sense that we use the term civilization. Civilization in this sense is attributed to both individual human beings and the groupings or collectivities of human beings. This was, as we have outlined, the original meaning of the concept when it was first introduced in the eighteenth century. In this sense, it is implied that there are 'civilized' and 'uncivilized' ways of doing things, or patterns of behaviour. Understood as such, civilization or being civilized signifies a quality or property of both man and groupings of men, which is approved, valued and exalted. Civilization is something good. It is a highly value-laden meaning.

Strictly speaking, this meaning of civilization can hardly be maintained. As Elias noted, 'there is almost nothing which cannot be done in a "civilized" or an "uncivilized" way'.[23] In this adjectival meaning of the concept of civilization, the 'civilizedness' and 'uncivilizedness' persist. They do not eliminate each other, though one of them (the civilizedness) is exalted. If there is a 'civilized' way of doing something, it necessitates that there is an 'uncivilized' way of doing it.

(ii) Civilization as a condition and process

The second meaning of the concept of civilization refers to a particular condition of men and societies, and also to a process the result of which is that particular condition, called civilization. In this sense, the term denotes a name for a process and condition or state of society. It is interrelated with the first meaning because it is the condition of civilization, or the result of the process of civilization, that allow men to attain 'civilized' behaviour, or the quality of being 'civilized'. Civilization as a condition implies that the society or men attained a particular condition at a particular point in time, a condition that men had not had before. Civilization as a process implies that the condition of civilization, which is itself a result of a process, is not finite. In other words the civilizing process is continuous. In this sense, civilization is conceived to be communal, that is, it is something which happens to a community.[24] Furthermore, as a process, it assumes the existence of a further condition, sometimes thought to be 'better' than the present one. It is in the light of this view that Collingwood treats civilization as a process of approximation to an ideal state.[25]

Civilization as a state or process is understood to be a general feature, applicable to, and attainable by, human societies, in fact all human societies. The view that civilization is a universal process or property, so common to the Enlightenment and post-Enlightenment authors, was best expressed by Guizot: 'For my own part, I am convinced that there is, in reality, a general destiny of humanity, a transmission of the aggregate of civilization; and, consequently, an universal history of civilization to be written.'[26] Civilization thus becomes a phase in the course of the life of human beings or societies; it was achieved throughout the history of mankind. Civilization is, in this sense, generally defined in contrast to some other condition(s) of humanity, experienced before the emergence of civilization such as barbarism, savagery or primitive condition. Here, we speak of civilization, not of civilizations. The word is understood in singular form. What this singular form of civilization, or the state of civilization, involves and how it is to be distinguished from other states of society are to be examined below.

(iii) Civilization as a collectivity

The third meaning of the concept of civilization refers to its plural form. By this meaning, we speak of civilizations, denoting that there are separate, distinct societies of human beings which have their own identifiable characteristics worthy of being called 'civilized' or 'civilization', as expressed when we talk of 'Western civilization', 'Chinese civilization', 'Islamic civilization' and so on. Braudel tells us that civilization (and culture) moved from the singular to the plural in the nineteenth century and the word 'civilization' began to be used in the plural in about 1819. The idea of the plurality of civilizations can, of course, be traced further back. Herder could be credited with implying it with his emphasis on the uniqueness and individuality of each people (*volk*) and on diversity as opposed to unity towards which everyone was heading in the age of the Enlightenment.[27] Civilizations in the plural form imply the renunciation of a civilization defined as an ideal, or as the ideal.

The plural conception of civilization does not only imply the existence of different, distinct societies in the state of civilization, but also different understandings of civilization as a state or property of society and man. One should therefore speak not of the progress of civilization in general, but of the progress or development of the separate European, Chinese, Egyptian or Muslim civilizations. With this meaning, civilization becomes a social entity with which a collectivity of people identify themselves or could be associated. It refers to a particular grouping of human beings and signifies one of the various forms of human identification. Civilization in this sense may be compared to other groupings of human beings and forms of human identification such as tribe, ethnicity, nation and so on.

The third meaning of the term civilization may be interrelated to the second and first ones. Firstly, we label each social entity or collectivity as a civilization assuming that something common enables us to speak as such, however they may have different conceptions of being civilized. Secondly, there have been some common traits that are historically traceable. Taken in this sense, of course, the question is the delineation of civilizations, which will be tackled in the next chapter. The remainder of this chapter is devoted to the account of civilization as a state, property, and process of human society (the first and second meanings outlined above).

Civilization and primitivity

When civilization is understood as a state or condition of human society, it implies that there may be other states or conditions of humanity. In other words, civilization as a state is defined in relation to other states that are sometimes described as pre-civilizational, or 'uncivilized', states. Seen as a general stage in the course of human history, civilization becomes a stage in human development and it is taken to be 'superior' to, or 'better' than, or 'improved' upon, the earlier stages. The earlier or pre-civilizational stages of human society are designated as the state of 'primitiveness', or 'savagery', or 'barbarism'. Such a view of the stages of human development seems to rely on an idea of progress as defined by Bury. According to this idea, civilization signifies a desirable and valuable state, whereas primitiveness, or savagery, or barbarism is equated with disdain and is devalued. Civilization is defined against 'primitivity' and it comes to have an ethical quality.

The view that mankind developed from an earlier 'primitive' condition to a later 'civilized' condition was so common among the historians, anthropologists, and ethnologists of the eighteenth and nineteenth centuries that it was almost taken as a presumption. For instance, while Condorcet (1795) outlined the rise of primitiveness through stages of animal husbandry and agriculture to alphabetic writing and ultimately the Enlightenment, Klemm recognized three stages: savagery (*wildheit*), tameness (*Zahmheit*) and freedom. According to Letourneau, all societies had their barbarous or savage infancy, out of which they have slowly and painfully evolved. Morgan divided all history into three main stages – savagery, barbarism and civilization – and correlated each with economic and intellectual achievements. Savagery was the period before pottery; barbarism, the ceramic age; and civilization began with writing.[28] Marx and Engels accepted this trilogy of the human past.[29] Tylor also agreed with this three-stage account: 'Development of culture, in great measure, corresponds with transition from savage through barbaric to civilized life.'[30] Although Tylor seems to be more careful by qualifying it 'in great measure' and refrains from making a final judgement, nonetheless he endorses the evolutionary course of savagery–barbarism–civilization.

The description of human history along a line of development means that those occupying a point on that line are more developed than those at the previous points. The idea that civilization, described as its own state of a society throughout its known history, represents a higher and better stage in both material and moral terms and that the pre-civilization state is, in Hobbes's famous words, 'nasty, brutish and short', existed in the classical period of Greek society. According to Moschion, who lived about the third century BC, it was due to 'Time – the begetter and nurturer of all things – that the Earth once barren, began to be ploughed by yoked oxen, towered cities arose, men built sheltering homes and turned their lives from savage ways to civilized'.[31] Considering that in all societies there is an accumulation of knowledge as a result of the socialization process, there can be nothing wrong with an idea of development or of progress in this sense.

However, when it is understood in Bury's definition, and we know that it has mostly been understood as such, then the idea of progress or development contains a view of ethical supremacy. As a result of this, civilization becomes ethically superior and more valued than other states. Moreover, other states of society are treated in degrading terms. The pejorative connotations of the words 'primitive', 'savage' and 'barbarian', which still occupy our language and culture, are clear indicators of this degradation.

From the eighteenth century onwards, many writers, in their depiction of in what sort of a condition mankind's earlier stage had been, went as far as to treat it as completely divested of any capability and value. Even Gibbon, the great Enlightenment historian, wrote that according to 'the discoveries of ancient and modern navigators and the domestic history, or the tradition of the most enlightened nations, the human savage [was] naked both in mind and body, and destitute of laws, of arts, of ideas, and almost of language'. He added that from this perhaps 'primitive and universal stage of man' he has gradually arisen up to measure the heavens.[32] Letourneau described it as 'border[ing] on animality'.[33] It was Sir John Lubbock who went to extremes in this respect. For him, the Andamanese have 'no sense of shame' and 'many of their habits are like those of beasts'. The Greenlanders have no religion, no worship and no ceremonies. The Iroquois have no religion, no word for God. Therefore, 'there can be no doubt that, as an almost universal rule, savages are cruel'.[34] Acknowledging that there are admirable moral standards in savage life, but seeing them as far looser and weaker than those of civilization, Tylor asserted: 'The general tenor of the evidence goes far to justify the view that on the whole the civilized man is not only wiser and more capable than the savage, but also better and happier, and the barbarian stands between.'[35] These remarks may be taken as an expression of the inherent self-righteousness that could exist in all societies and men. Or, they may be seen as the expression of how the Europeans, who have become sure of their superiority from the Age of Exploration onwards, perceived others. However it is understood, in all these views, there is an idea of progress, civilization being a result of it, and still ongoing. Primitiveness or savagery, whatever term is used, signifies what the earliest condition of man had been like.

Perhaps it should here be noted that there are those who argue for a 'degeneration' view according to which the earliest phase of human history was not one of savagery or primitiveness, but to use the familiar term, one of civilization. The most well-known version of this idea is the Hebrew myth of 'The Fall' described in the book of Genesis. However, the idea of a primeval 'Golden Age', that is, a period of perfect civilization, can be traced back to the Sumerians. For them, its most significant feature was freedom from fear. In the words of a Sumerian poet:

> Once upon a time there was no snake, there was no scorpion,
> There was no hyena, there was no lion,
> There was no wild dog, no wolf,

There was no fear, no terror,
Man had no rival.[36]

The view of a primeval 'Golden Age' or degeneration theory holds that man is civilized by birth, by creation. He was not 'naked both in body and mind' and not in an abject condition. Religions and theology support this description of the earliest state of humanity. In fact, both the *Old Testament* and the *Koran* do not seem to depict mankind's earliest phase as one of primitiveness. According to the Old Testament, Man, in the beginning, had the 'tree of the life and the tree of the knowledge of good and evil'. The Koran explicitly states that God 'taught Adam the names, all of them'. Man had the knowledge in the beginning, according to the Koran as well.[37] The degeneration view then holds that the primitive condition is neither the earliest period of humanity nor a universal state. Primitiveness is a condition of degeneration and those people who are found, or said, to be primitive in the contemporary period or before are degenerated forms of the earlier civilized societies by birth. Yet in the literature, it has been the prevalent view that the earliest stage of human history had been one of primitivity.

How do we know that mankind experienced an earlier primitive stage? Archaeological findings of the remains of earlier societies or men are of course the principal source. Another source is to compare the tribes or small societies which are said to have a primitive condition and thus to consider their state to be similar to the earliest state of human beings. Yet, the limitations of these sources are evident as there are limitations to historical knowledge and knowledge itself as well. A particular finding or a particular observation is no grounds for a universalization of the stages of all human beings. It is true, one cannot prove the existence of a primitive period of humanity through comparing and assuming the contemporary tribes to be an example of mankind's earliest phase. The limitations of the comparative method have long been acknowledged.[38]

Nonetheless, these sources can legitimately be used to obtain knowledge about the various states of human society at various periods, as the limitations have always existed in our knowledge. Therefore, the problem is not the limitations of sources, but what is made of sources. The progressionists made a universal developmental course of human history out of these sources not because the sources were common to all societies and peoples, but just because they constructed the sources in the light of the idea of progress and a common human nature. They thus ended up in a universalism as if all peoples had to experience the same course of history which they constructed. For example, while Tylor speaks of treating 'mankind as homogenous in nature',[39] Morgan, as a rule, adheres to the principle that 'the experience of mankind has run in nearly uniform channels'.[40] Thus, it is not the archaeological findings of societies in the past or current anthropological observations that provide a universalistic course of the development of civilization, but the idea of progress which developed in the modern period. Then, it can be seen that such an understanding of civilizational development unites with the objectivist account of universal history examined in the previous chapter.

Nevertheless, it makes sense, in my mind, to speak of the 'development' or 'progress' of civilization as a state of society. As I have already said, the accumulation of knowledge and the socialization process allow us to express it. This statement in no way leads us to the view that humanity as a whole has gone through 'uniform channels' from 'savagery' via 'barbarity' to 'civilization'. However, the existence of other states of societies, which are to be characterized as 'uncivilized', or 'primitive', or 'savage' could be historically asserted and shown. What cannot be historically, and logically, asserted and shown is the 'cruelty' of the earlier states of men, as cruelty does still exist in men. The idea of development or progress of civilization does not justify one despising the pre-civilizational states. Unless one adopts an objectivist view as defined in this work, one can hardly make an ethical point out of civilization. One can legitimately speak of civilization in, what I have called its 'technical' sense as a state of society, which is but the result of what men have produced and inherited. Tylor has, as already said, come close to it, despite his ideas which fall in the objectivist position. What remains now is to distinguish civilization as a state of society from the others.

Landmarks of civilization

How could the distinction be made between, to use the familiar terms, 'civilized' and 'primitive' states of man? In other words, what are the features or characteristics of civilization that mark it off from the non-civilization? There have been a number of criteria for civilization.

The first criterion proposed, especially by the archaeologists, was the absence or presence, high or low development, of the industrial techniques especially metal-working. In other words, the dawn of civilization was equated with the appearance of metal tools.[41]

Secondly, it has been suggested that the invention of writing differentiates civilization from the primitive condition just as it is taken as the dating of history from prehistory. As I have already said, Morgan equated civilization with writing. Eckhardt and Wright agree with Morgan in defining primitive people. For Eckhardt, primitive people refers to 'primarily non-literate people, that is, people in whose lives neither reading nor writing played any real role. This distinguishes them from "civilized" people, some of whom learned to read and write about five thousand years ago'. Wright defines primitive people as 'human beings that live in self-determining communities which do not use writing'.[42] One can find other examples, but there is no need for that. The view that equates the emergence of civilization with the use of writing is a prevalent one. The rationale behind this view is that writing makes it possible to keep the records. Yet, possession of script can hardly be taken as a distinctive criterion for civilization for it is not the only means of keeping records. Human memory (oral transfer from generation to generation) provides another means of record-keeping. For instance, the bulk of the Hindu scriptures have been preserved and transmitted for centuries by memorization. Moreover, Bagby points out that at least one society, which is usually characterized as civilized, namely the Andean, was non-literate.[43]

Childe, proposing the third criterion, argued that civilization emerged with what he called the 'urban revolution'.[44] According to this idea, civilization came into being when men gathered in cities and began to specialize. Occupational specialization resulted in the rapid improvement of skills, and the invention of new and better tools. Consequently, it is held, we see a sudden elaboration of the material, intellectual and artistic aspects of society to a level of complexity and refinement that is called 'civilized'. Childe's argument is shared by many. According to Engels, the division of labour was the most important factor in the emergence of civilization. The first division of labour emerged in the middle stage of barbarism between the pastoral peoples and the backward tribes without herds. The upper stage of barbarism introduced a further division of labour between agriculture and handicrafts resulting in the production of commodities for exchange. Civilization strengthened all these divisions of labour, particularly, by intensifying the contrast between town and country and added a third division of labour: it created a class that took no part in production, but exclusively engaged in exchanging products, namely the merchants or traders.[45] While Redfield equated civilization with the rise of cities on the grounds that 'the administrative elite', 'the literate priest' and 'the specialized artisan' made their first appearance in the cities, Frankfort endorses the view that town is where society is in a condition of civilization.[46]

The association of the emergence of civilization with the rise of cities, and the rise of cities with the beginning and intensification of the division of labour or social diversity has been the most commonly held view. Challenging the rigidity of these views, Bagby argues that the clue lies in the etymology of the word 'civilization'. He too equates civilization with the city life, meaning that the social life of community is centred in cities when it is in a state of civilization. Yet, he rejects the significance of the far-reaching division of labour, together with the significance of population size and literacy, for classifying a society as being civilized.[47] What matters, for Bagby, seems to be whether there are cities, agglomerations of dwellings, the majority of whose inhabitants are not engaged in producing food, in the sense that the social life of a community is centred around them. To my understanding, Bagby does not refute the necessity of the division of labour, but he appears to be of the view that it is not all-important.

A fourth criterion has been suggested as being the establishment of settlements. Those in favour of this view argue that civilization came into being not when men began to live in cities, but when they began to have a settled life as opposed to hunting-gathering or engaging in nomadic life patterns. In this view, again, the socialization and social diversity have a crucial role, but they are not wholly equated with the cities. Settlements suffice for these characteristics (specialization and diversity) to develop, and first settlements were, it is held, agricultural communities. McNeill makes a typical account of this view. For him, occupational specialization, diversity of personal life experience, and persistent invention and new creation, which result from the tension between innovation and routine, are obvious characteristics of civilization.[48] Behind the emergence of all these characteristics lies the fact that the regular and abundant harvest made farmers capable of supporting others who did not have to support themselves and were able to engage in administrative or priestly affairs. This paved the way for social diversity and highly skilled societies that were called civilized between 4000 and 3000 BC in the lower Tigris-Euphrates valley.[49]

A variety of other distinguishing characteristics and features of civilization have been put forward by various authors. According to Toynbee, the difference between primitive societies and civilized societies is dynamism, or the direction of mimesis. In primitive societies, mimesis is directed backward towards the past whereas in civilization it is directed forward towards the future.[50] In Spalding's view, the differentia of civilized society is that, unlike primitive society, it is not susceptible to interpretation in biological terms. The primitive thinks that the maintenance of group or clan is important while, for civilized man, the survival of the individual is important.[51] Lord Raglan and Eliot argue that civilization originates in ritual and religion.[52]

Some authors proposed a combination of various features which were taken as the decisive factor in the transformation of primitive societies into civilizations. As early as the first half of the nineteenth century, Mill emphasized the unity of, in his terms, the 'ingredients of civilization' and described a range of characteristics from the magnitude of population, via the city, industry and exchange to social mobility and cooperation, and organized, lawful government as the distinguishing characteristics of civilization.[53] Quigley, nearly one and a half centuries after Mill, proposed a combination of production, writing and city-life as the distinguishing feature of civilization.[54] Considering a combination of various characteristics and developments as the distinguishing features of civilization may come to be a 'better', more 'adequate' and more 'historical' explanation than the one based on a single-cause search. When such an argument is advanced there is the urge to know which one is 'primary' and 'decisive'. Yet thinking of the historicalness of men, it seems to me that the 'primary' or 'decisive' factors, if there are such factors, vary according to specific communities.

Following Collingwood, I go further. It is hardly possible to make an absolute or sharp distinction between the so-called primitive and civilized conditions of a community. First, to make such a distinction one should adopt an unhistorical view and stop the civilizing process at a point. If civilization is a condition of society, it means that it is a result of a process, that is, a civilizing process. It is not only a result of a process but also operates within a process. If the civilizing process is assumed to have started some time ago, then, it follows that some other process had ended before. Historical and social continuity does not allow us to make such an absolute dichotomy. Moreover, if it is assumed to be possible, then the word 'civilize' and process cease to have any meaning. In Collingwood's words, 'a process has no absolute beginning and no absolute ending'.[55] The question is not if a society is civilized or barbarous, but civilized or barbarous in relation to what.

Second, civilization as a social process becomes meaningful, or is understood, within the community in which it takes place, as I have already hinted when I said that the 'primary' or 'decisive' factor might vary according to specific communities. Here, we go on to the plurality of civilization, and begin to speak of civilizations. It follows that every civilization has its own standard of civilized life. Then, it becomes impossible to make a distinction between primitiveness and civilization, for every society is civilized, otherwise it would not be a society. Now, the question is in what way a society is civilized. And, again in Collingwood's words, 'from the point of view of any one civilization, any other is merely one of the innumerable forms of barbarism'[56] as the Europeans and Chinese may think of each other as barbarous. This is historical relativism as expounded by, say, Spengler.

As I have already stated, the plurality of civilizations does not deny the existence of a general concept of civilization. As long as we speak of 'civilizations', there will be 'civilization'. Furthermore, as I mentioned earlier, there are historical commonalities. For instance, the invention of the plough is said to be a mark of civilization. The way it is made or used may be taken as a characteristic of a civilization. Similarly, if the invention of writing is seen as a mark of civilization, the Latin, Arabic and Chinese scripts could mark different civilizations. That is where we come to civilization being a social entity or collectivity. In other words, we do not just speak of civilization in the singular, but civilization*s* in the plural. Understood as such, then, civilization becomes a unit of identity for human beings and a unit of analysis for the students of history. This constitutes the subject matter of the next chapter.

5
Civilizations or Realities of the Extreme *Longue Dureé*

As I have already set out in the previous chapter, civilization in its plural form denotes a particular type of societal entity or unit. Compared to the more familiar units such as nation, ethnic group, tribe and so on, civilizations are large-scale collective identifications. Viewed this way, of course, the question is how to distinguish or delimit multiple civilizations. Unlike its meaning as a state of society, in this meaning it can be taken as a unit of analysis and some historians such as Spengler, Toynbee, Hodgson and McNeill argue that civilizations are 'proper' units of analysis for history. Saying that civilizations are the proper units of analysis for historical study is not the same as saying that other collective (social) entities and identifications are improper units. It is my view that the multi-faceted nature of social phenomena does not allow only one unit to be taken at the expense of the others. Civilizations are, I have said, one form of social identifications. Then, what is a social (collective) identity? And first of all, what is meant by identity?

Identity and identification process

It has already been emphasized that identity and existence are co-attributes or processes. Identity can then be said to be a definition or qualification, by which we come to know or describe something. Someone or something that we can describe by identity, or with which identity is associated may be named as *the identifier*; and in turn the thing that defines and describes the identifier could be said to be *the identified*. For example, individuals are identifiers and 'society' is the identified, and the result is what we call 'social identity'. It can fairly be argued that, despite the seemingly opposed connotations, identity and distinction are closely interdependent.

Identity and difference/other

Any identification therefore requires a distinction just as any distinction necessitates some identification. That the identification process necessitates a process of distinction means a unit of identity requires some difference or other. Indeed, years ago, Levi-Strauss stressed that there was a twin relationship between the self and the other. With Derrida, we can now say that every identity exists together with its difference; there can be no collective social identification without its own difference or other.[1] Identity and difference exist in all identification processes and units of identity. For any unit of identity and identification process, there is a need for the other. This can be shown both logically and historically.

Logically this is so because to identify something means to differentiate it and similarly to identify yourself with some group requires you to distance yourself from some other. For example, somebody who identifies himself with a family does not identify with another one. However, the family/families with which no identification has been made does/do exist. If the families with which a particular person does not establish identification did not exist, then, there would be no need for identification with the particular family. Similarly, someone who is, for instance, a Turk is also the one who is not an English or French or Greek. If all nations other than Turks would disappear then there would be no need for one to identify himself with the Turkish nation. The Turk exists because the English exists. Any identification therefore requires a distinction.

That all forms of identity involve some sort of difference may be expressed the other way round: all difference involves identity and any distinction necessitates some identification. If nothing is identified, then, no distinction can be achieved and vice versa. The identity of something depends upon the existence of something else. The need for the other or difference in defining identity thus comes as a logical condition.

Furthermore, the existence of the other/difference in the collective identity formation is not just a logical necessity, but it also appears to be a historical fact. Historically, we see that the identities of all societies have been defined through their difference from other societies. In the identification process, differences from others have sometimes been more defining than the commonly shared characteristics. Thucydides tells us that the Hellenes are identified through their difference from the Persians. According to Thucydides, before the Trojan War there was no identification of being 'Hellas' or 'Hellenes' in Greece. Homer did not call them by the name of 'Hellenes'. He did not even use the term 'barbarian', probably because the Hellenes had not yet been marked off from the rest of the world by one distinctive appellation.[2] Similarly, English and French mutually determined each other. It has been a generally agreed view that in the definition and formation of the English and French national identities, the Hundred Years' War was one of the most significant factors. It was this continuous conflict, Trevelyan wrote, which supplied England with 'strong national self-consciousness; great memories and traditions; a belief in the island qualities'.[3] It is a commonly held view that the Ottoman power in the East had been a prominent factor in the shaping of the modern European identity.[4]

The role of the other in identity formation may be followed through the names of identity units. The units of identity are usually named by others or named upon the encounter with others. As already seen this is what happened in the case of the 'Hellenes'. The same goes for the 'English'. As everybody knows, the word 'English' is not English, but Latin. The people called 'English' were not named by the English themselves, they were named by the Romans (Latins). Similarly, the root word of the word 'Turk' is not Turkish, but comes from Chinese. The Finns were named by the Swedish and the Kurds were named by the Turks. There is no need to extend these examples. The existence of the other seems to be a requirement for the definition of identity. If nothing is identified, then, no distinction can be achieved and vice versa. The identity of something depends upon the existence of something else.

Identity and the individual

Man identifies himself with something else, for example, family, sex, group, nation and so on in order to become what he is. It has been argued that at the most fundamental level; identity results from human vulnerability. In order to have psychological security, every individual is said to possess 'an inherent drive to internalize' – that is, to identify with – the behaviour, mores, values and attitudes of those in his or her social environment. Moreover, so it is held, every human being possesses 'an inherent drive to enhance and to protect the identifications he or she has made'.[5] The lengthy and vulnerable infancy and childhood of human beings may be taken as the basis of the 'inherent drive' for identification with others to achieve psychological security and, perhaps, physical security. Whether the drive for identification is inherent or not, identity and identification are a concomitant part of human existence.

Identification comprises the processes of inclusion and exclusion. On the one hand, we see that the identifier internalizes or associates with the values, behaviours, attitudes, symbols and myths of the identified, and on the other hand, an externalization of, or dissociation from, the values, myths, symbols, attitudes and mores of the non-identified. Human existence is a social formation and the identification process is not a process which an individual undergoes by himself in an isolated condition. All identifications of human beings are then social.

Identification is an on-going process. This means both the enhancement of the existing identifications and the establishment of new identifications. In other words, human beings have multiple identities. It could, I think, be possible to advance a threefold classification of the identifications of human beings. Everyone has different traits and characteristics, yet people have similar features as well. Thus, separate and common identifications are possible. First, there is a common identification of everybody, for all persons share some traits with all others, that is the universal characteristics of the species. This could be regarded as the identification of all human beings as distinct from non-humans. Second, as all persons share some characteristics with some others, those characteristics define those persons as members of a particular group, leading to what may be called group identifications. Any group identification accordingly involves the existence or establishment of commonalities to form the group and of differences to distinguish it from the others. Third, a person has some traits which he or she shares with no one else, constituting his individual personality or idiosyncratic characteristics, making the personal or individual identity.

If the identification basically arises from the human infant's need for physical and psychological security, this means that, as already stated, the security need, that is, survival, forms the basis of identity. Behind this need for security lies the vulnerable character of human beings. As human beings may have various characteristics and needs, or they may develop and obtain them in time, then it could be asserted that human beings may have, or may achieve in time, multiple identities. They can identify themselves with various groups according to their characteristics and needs. The scope, intensity and number of these identifications will vary with the degree and strength of these needs and traits, and with time and place as well. As Smith rightly puts it, there is nothing to prevent individuals from identifying with Flanders, Belgium and Europe simultaneously, and displaying each allegiance in the appropriate context; or from feeling they are Yoruban, Nigerian and African in concentric circles of loyalty and belonging.[6]

It follows that the personal identity of an individual is formed by his collective identifications. Personal identity is indeed social. When the child internalizes the patterns of behaviour of, say, his parents, he internalizes a behavioural pattern on which the social environment has already made an impact. Even the idiosyncratic characteristics of an individual do not just emerge from a biological or an intra-psychological process, independent of the social milieu in which he or she is socialized. Moreover, it has been said that for most psychologists, from Freud to Mead, personality is a social construct and the result of social interaction. Bloom, I think rightly, adds that to a lesser or greater degree, all identifications are social and shared. The identifier might be an individual, but the identified is always social.[7] What one is can only be intelligible in the social network in which one is a member. The individual self is not something detachable from one's relationship with the others. What one is may depend upon what one feels and thinks and what one feels and thinks is not independent of the prevailing feeling and thought in the society of which one is a member.

Having said that identity emerges out of the needs and common traits and characteristics of human beings, I could further propose that societal (or group) identification is evoked and enhanced if, (a) the group provides individuals with security in the face of external threat, and/or (b) the group is beneficent towards the (would-be) members. The existence of an external threat causes human beings to make identifications with the others around, just as the vulnerability of the human infant against the environment leads him/her to make identification with the parents. Through the existence of external groups – that is, 'the other' – a group is distinguished and identified. When external groups are perceived as a threat, it is highly likely that the distinction increases and the group identification is enhanced. It is also natural to expect that the cohesion of a collective identity unit increases if the members of the said collectivity consider it good for themselves. It can then rightly be asserted that, besides the provision of security in the face of external threat, identity consciousness is also evoked if the collectivity in concern is beneficent towards the (would-be) members. But, what is a collectivity or collective identity?

Collective identity

As a working criterion, I adopt the view that a collective/social identity could be defined, first by common 'objective' elements shared by some people, such as language, common descent, history, customs, institutions, religion, myths, symbols, style and so on, and second by the 'subjective' self-identification of people in concern, meaning that people are conscious of their commonalities and that they consciously identify themselves with the said collectivity.[8]

A collective identity begins to appear when the members of the said collectivity internalize the objective elements. The socialization process which the human infant goes through in every society largely carries out the internalization of the elements of a collective identity. The fact that the human infant internalizes many of the values and elements of the society into which he or she is born accounts for the continuity of the said collective identity. This however by no means implies that the internalization, and thus identification, process is confined to the period of human infancy. It is, as already said, an ongoing and permanent process throughout the lifetime of an individual.

The salient features of the collective identity in question may be formed according to the emphasis on or priority of a particular element. The element of religion, for instance, was emphasized in 'Europe' throughout the medieval period so that Europe and the Europeans were basically regarded as synonymous with Christendom and the Christians. The Christian remained a significant defining epithet for the European until well into the late eighteenth century.[9] In the modern period, it could be said that nationality and 'civilization' or 'being civilized' have replaced the saliency of religion in the European identity. Today, the 'Copenhagen criteria' or more properly 'Copenhagen political criteria' seem to be the determining element of being European. Similarly the salient feature of nationality also changes according to where the stress is laid. When territory and mutual rights and obligations are prioritized in the account of national identity, we speak of civic nationality and when common descent and language are prioritized then it is ethnic nationality.

The process of the internalization of the objective elements shared by the members of a collectivity leads us to the subjective elements of the collective identity. To speak of a collective social identity, besides shared characteristics, the members of the collectivity should have some sort of subjective consciousness of belonging to that collectivity. In other words, we need to see if human beings identify themselves with a social entity that we consider to be a unit of identity. Apparently, this is an empirical question. However, it must be stated that the distinction between 'objective' elements and 'subjective' identification, put as a working definition, does not mean that the two processes are independent of each other. The very existence of 'objective' common elements indeed reveals that there is a self-identification of human beings with each other and consequently with all the others having the same elements or characteristics. If people have nothing in common, how could they be expected to identify with each other? The question is not, then, whether there is an identity consciousness, but how cohesive or how strong it is. The question of degree is in turn something that depends upon the specific units of identity and historical conditions.

The collective social identifications that an individual could have range from the smallest collective unit, for instance, family, to the all-embracing one, for example, humanity. Smith lists six of them. The first one is the category of gender that is universal and pervasive. The very universality and all-encompassing nature of gender classification makes it a less cohesive and potent base for collective identification and mobilization. Second, the category of space or territory, which has a more cohesive quality than gender identification. The third collective identity is based upon, in Smith's words, the 'sphere of production and exchange', namely, the category of social class. The class identity is, like gender identity, a less cohesive identification as it is dispersed by territory, religion and ethnic groups. Fourth comes religious identity deriving from the spheres of communication and socialization, and based upon values, symbols, myths and traditions, and often being codified in custom and ritual. The fifth collective identity, linked with religion, is the ethnic identity. An ethnic community, or an *ethnié*, for Smith, has six main attributes: a collective proper name, a myth of common ancestry, shared historical memories, one or more differentiating elements of common culture, an association with a specific 'homeland', and a sense of solidarity for significant sectors of the population. The last collective identity in Smith's list is national identity whose fundamental features according to the two models of the nation (that is, the one civic and territorial, the other ethnic and genealogical) may be outlined as follows: a historic territory or homeland; common myths and historical memories; a common mass public culture including vernacular languages and shared customs and traditions; common legal rights and duties for all members; and a common economy with territorial mobility for members.[10] Smith's list is of course in no way exhaustive of the existing or possible types of social identifications. Yet, I have taken it because it could give us some clues to detect certain points about social identifications.

First, as I have already hinted, all collective identifications are social in the sense that man is *per se* a social creature. Parallel to this, they are historical in the sense that they are formed in the course of time.

Identifications, being social, involve collectivities of human beings. As already stated, even the individual identity takes place in, and is shaped by, a collectivity. A social identity is the internalization of some elements by some individuals. Man, being a social being, as far as we know him, has always had a kind of feeling for collectivity or group consciousness, as emphasized by Kohn.[11] The extent of this group feeling and the size of the collectivity it covers may vary. The collectivity of a social identification may even grow to the extent that it can gain an all-embracing character as it is in gender identification. A family or a simple group composed of three or four people camping together is, for example, a small collectivity of social identifications. Communities, nations, *ethniés* are prevalent and familiar collectivities. Large collectivities such as religions and civilizations also constitute units of collective social identification.

Being social, collective identifications are historical, too. The historicalness of a social identity involves not only the case that it takes place in the historical process, but also it means that the extent and degree of their collectivity, their cohesiveness and impact upon human beings can change throughout history. At a particular period in time, for example, a particular form of identity may have a greater cohesiveness and consequently influence individuals, as was the case with religion before the modern period in Europe. In another age, another form of identity may gain strength and exert control upon individuals, as in the case of national identity with the modern age in Europe. A form of identity in time may even get near-universal. The point has been made that the national identity has been so predominant in the contemporary world that 'to be without nationality is to be perceived as almost without identity'.[12]

The second point to be noted about social identifications is that two classes of them can be defined on the basis of how human beings achieve their identity. One refers to the identities chosen by individuals with their own will and the second one denotes those into which individuals are born. Examples of the first are interest groups, political parties, clubs and associations; examples of the second are family, *ethnié*, society, nationality and civilization. Those identifications which are not chosen by individuals are in the beginning exclusive as individuals are born into them. One can only be born into a family, a society and a civilization, and are thus excluded from the others. However, this involuntary and exclusive character of some social identifications is not absolute and continuous in the lifespan of individuals. The very sociality and historicalness of social identity allows men to change their identificational collectivity. In time, man could change his family, his nationality, his religion and so forth. Paradoxically, it is that very sociality of the social identity that, via the progress of socialization, makes it extremely difficult for individuals to change their social identifications such as religion, culture and civilization.

A third point is the extent and degree of social identifications in making cohesive units and mobilizing individuals for common action. As I have already stated, this could vary in the course of historical process. Yet, some propositions may be advanced. Smith says that the very universality and all-encompassing nature of gender classification makes it less cohesive and a potent base for collective identification and mobilization. Following him, it could be proposed that the wider an identity is, the less likely it is to be grounds for a cohesive identification and mobilization for a common action. That is why civilizations are less cohesive and potent than, say, nation-states. The need from which a social identity emerges may be a determinant of its cohesion and capacity for collective action. It may increase according to urgency and necessity. Another factor, already mentioned, can be given as the existence of an external threat. One other may be recalled. The degree of cohesion and mobilization of a collectivity may depend upon how beneficial the identification is towards individuals.

As a last point, it could be recorded that a social identity, once established, may lead to, or enhance, other identifications. This is a result of the dynamic and ongoing nature of the process of identification. It is a commonplace that territorial identity largely determines one's societal or national identifications. Conversion to a particular religion may lead to a change of communal identity. Similarly, identifications of class, nation, *ethnié* and religion could increase or decrease the cohesiveness of each other. Having made these observations on identity, now we can examine civilization as a unit of identity.

Civilizational identity

Historically speaking, civilizations as social identifications have been large-scale collectivities when compared to other collective units of identity. They are large-scale entities in two senses: in their spatial coverage and temporal extensions. Civilizations are wider and broader and more durable and long-lived than other collective identifications in human history. This view is best expressed by Braudel. 'Civilizations are,' wrote Braudel, 'realities of the extreme *longue dureé.*' Civilizations are not 'mortal', at least, in comparison to human lives. They exceed in longevity any other collective reality and, in space, go beyond the frontiers of the specific societies. They are 'the longest story of all' and can be approached 'only in the long term'.[13] Toynbee earlier expressed the same points in defining civilizations as 'societies' which are wider in space and time than national states or any other political communities and short of embracing the whole of mankind and covering the whole habitable or navigable surface of the Earth.[14] While Hodgson defined a civilization as 'a relatively extensive grouping of interrelated cultures', McNeill included in it 'the totality of different and diverse groups'.[15] Recently, following Toynbee, Huntington has defined a civilization as 'the highest cultural grouping of people and the broadest level of cultural identity people have short of that which distinguishes humans from other species'.[16] Understood as such, we can say that civilization as a social unit/identity takes place in between what may be taken as the more usual social unit (e.g. a community or nation-state) and all-embracing unit/identity within the identification continuum of the present study. Being large-scale entities, civilizations thus incorporate a multiplicity of other social collectivities or group identifications. Of course, it makes them have a low degree of cohesion and potency. It should here be pointed out that large-scale characterization of civilization is just the result of historical observation. It entails no logical necessity.

Closely related to the view that civilizations are large-scale and long-lived collectivities is the idea that they are relatively self-sufficient and self-comprehending societies. The most ardent proponent of this idea is Spengler and one can add Toynbee. Yet, the idea is implicitly expressed by many writers. For Spengler, civilizations or, in his terms, cultures are the prime phenomena of all past and future world history. They have their own ideas, lives and death, their own possibilities of self-expression. Each civilization possesses its own sculpture, painting, physics, number and mathematics.[17] Toynbee considers civilization to be an 'intelligible field of historical study' and as institutions that 'comprehend without being comprehended by others'.[18] The basic idea behind this view is that a small or individual society can hardly be understood and comprehended when the analysis is confined to its boundaries because it is not self-sufficient and free from external influences and impulses. Accordingly, so it is held, a larger entity comprises more possibilities for self-sufficiency and self-comprehension. Ranke, who did not have a conception of separate civilizations, seems to have in mind this idea when taking Latin and Germanic peoples as a unit. Similarly, Collingwood, when explicating 'the idea of history', relied on 'the modern European idea of history', implying that 'Europe' could be a more 'intelligible' unit than, say, 'England'.[19] Hodgson and McNeill, as already stated, recognize the heuristic value of civilizations in examining world history, though both of them argued that mankind was the only ultimately tenable field for historical inquiry. The idea that civilizations are relatively self-sufficient collective identifications has, as seen, been quite prevalent among historians, even those who did not adopt a civilizational view of history.

The self-sufficiency and self-comprehension of civilizations can, however, only be argued in relative terms. The Spenglerian idea, according to which civilizations are conceived to be self-sufficient in terms of being free from external impulses and interactions, can hardly be maintained. There have always been encounters and exchanges among civilizations. Historically, in various regions of the world, we see different civilizations in interaction with each other.

As already mentioned, historians like Hodgson and McNeill consistently showed that there had been interactions and exchanges among different civilizations since the very beginning of the so-called civilized societies. Long-distance trade occurred when the river valley civilizations of Mesopotamia and Egypt began to import goods like metal and timber across quite considerable distances from the barbarian lands. Inter-civilizational trade, too, was very old. Mesopotamian commercial contacts with India dated back to the third millennium BC. Indirect and far more tenuous contacts between Mesopotamia and China started a few hundred years later, though caravans only began to move more or less regularly across Central Asia in about 100 BC. Nevertheless, with the passage of time the scale and range of trade exchanges within Eurasia expanded into Africa and then, after AD 1500, began to embrace all the inhabited Earth. The degree of the inter-civilizational contacts, for Hodgson and McNeill, was high enough to consider the Afro-Eurasian zone as an intelligible field.[20]

It is not, then, easy to view civilizations as self-sufficient and self-comprehending entities. If civilizations are to be taken as self-sufficient and self-comprehending, we need to ask, self-sufficient in what respects and self-comprehending in which aspects and limits. Civilizations can then only be regarded as relatively self-sufficient and self-comprehending. The same relativity goes for large-scale and long-lived characterization, in other words, civilizations are large-scale and long-lived relative to the other known collectivities of human life. We can, therefore, sum up that civilizations are long-lived, large-scale, self-sufficient and intelligible collectivities in comparison to other collectivities.

Delimitation of civilizations

Having examined the basic nature of civilizations, the question comes to the delineation of different civilizations. What could be the distinguishing elements? What would be the criteria for the delimitation of civilizational identities? In line with the general definition of a unit of collective identity advanced earlier, I adopt the view that a civilization could be defined, first by common 'objective' elements shared by some people such as common descent, language, history, customs, institutions, religion, style and so on; then, second, by the subjective self-identification of people in concern, meaning that people are conscious of their commonalities and that they consciously identify themselves with the said civilization. Let us consider those 'objective' elements.

Objective elements

Common descent

It might be argued that a civilization is formed by a collectivity of people who are supposed to come from a common descent or claim that they have a common ancestry. Though this view seems, to us, so obvious as to be unsustainable, when united with racial ideology, it goes as far as to assert that only those coming from a certain ancestor can form a civilization. De Gobineau, for instance, declared that 'all civilizations derive from the white race, none can exist without its help'.[21]

Alas, it is so obvious that a common descent or ethnic stock cannot be taken as a basis for a civilization. Isocrates knew it. In his *Panegyricus*, he wrote: 'And so far has our city distanced the rest of the world in thought and in speech that her pupils have become the teachers of the rest of the world; and she has brought it about that the cause "Hellenes" is applied rather to those who share our culture than to those a common blood.'[22] For Isocrates, then, not common blood but shared culture was the defining element of being 'Hellenes'. Herodotus, on the other hand, counts 'common blood' among the elements that unite the Athenians and Spartans. The Athenians reassured the Spartans that they would not betray them to the Persians:

> For there are many powerful considerations that forbid us to do so, even if we were inclined. First and chief, the images and dwellings of the gods, burnt and laid ruins: this we must needs avenge to the utmost of our power, rather than make terms with the man who has perpetrated such deeds. Secondly, the Grecian race being of the same

> blood and the same language, and the temples of the gods and sacrifices in common; and our similar customs; for the Athenians to become betrayers of these would not be well.[23]

As can be seen, for the Athenians, too, being of the same blood was not the only element distinguishing the Greeks from the Persians and other non-Greeks. Besides common descent, images and dwellings of the gods (common symbols and temples), burnt and laid ruins (shared experience), common language, religion and similar customs are counted as the commonalities the Greeks shared.

To those remarks by Isocrates and Herodotus, one can further add that historical research demonstrates peoples, having a supposed common descent, scattered in more than one civilization (Indo-European peoples being the identifiers of, at least, two or three distinct civilizations – Western, Indian and Islamic); and it also demonstrates civilizations with more than one ethnic stock (Islamic civilization comprising Arabs, Turks, Persians, to name but a few). It might be argued that the fact of your being born into a family from a particular ethnic stock could influence your social identities. However, this is not equal to saying that people from a common ancestry could constitute a civilization. Yet, the myth of a common descent, if sustained, could enhance civilizational identifications.

Language

It is said that separate civilizations possess distinct languages. Language does not, however, form a significantly distinctive element of civilizations as it changes with time and place. Moreover, civilizations are usually bilingual or multilingual. For example, Sumerian and Semitic in earlier Mesopotamia, Greek and Latin in Hellenism, Western Europe and Islam with many languages.

Nonetheless, each civilization may have a sort of 'supra-language', or *lingua franca*, throughout its geographical and human extensions – a 'civilizational' language or a language of power and elite. Kroeber gives some examples: Mandarin, Sanskrit, the Greek Koine, Latin in Western Christendom, Arabic in Islam, Great Russian in Soviet Civilization.[24] The existence of a supra-language or civilizational *lingua franca* may enhance civilizational identity. This is reasonable. Nevertheless, the changing nature of language and the multilingual characteristic of civilizations prevent language from being the distinctive element.

Religion

Religion is taken as a distinguishing element of a civilization on the grounds that civilizations are accompanied by religions or some kind of worship. Northrop, for example, considers the characteristic of religion to be one of the distinguishing elements of what he calls the 'Eastern civilization' and 'Western civilization'. The religions of the Eastern civilization (Hinduism, Buddhism, Taoism and Confucianism), unlike those of the Western civilization, are not theistic. In other words, the divine is not identified with an immortal, non-transitory factor in the nature of things, which is determinate in character. The divine is indeterminate, and these or specific determinate properties designate what it is not, not what it is. Moreover, the religions of the Eastern civilization are not pantheistic, that is, the divine is not identified with the totality of the universe. None of them has, or worships, a prophet.[25]

As most civilizations comprise some form of religion, it has been a significant element. Some scholars defined civilizations on the basis of religion. Toynbee, in his early years, saw religion as one of the most distinctive elements of a civilization. According to Dawson, 'the great religions are the foundations on which the great civilizations rest'. 'Of all the objective elements which define civilizations', Huntington argues, 'the most important is usually religion.' Eliot even goes further and asserts that no civilization could come into being without a religion.[26]

We may have our doubts about Eliot's assertion; yet, we cannot deny the significance of religion in understanding civilizations, as we know them. Some civilizations have institutionalized religions such as Christianity in both its Catholic and Orthodox versions, Islam, Brahmanism and Confucianism, whereas some civilizations (for instance, the early Near Eastern civilizations) do not have organized religions, but local forms of worship. Some civilizations are distinctively defined or established by their religion as in the case of the Islamic civilization. Yet there have been those civilizations like the Indian and Chinese civilizations, which include more than one religion.

Territory and geography

I have earlier said that a territorial region, or geography in general, constitutes one of the major factors in the emergence of civilization as a state of society. Hence, it would seem reasonable to expect that differences in territory and geography could lead to civilizational differences. It is true in the sense that geographical conditions affect the way human beings do things. Many of the characteristics of civilizations depend on the constraints or advantages of their geographical situation, as stated by Braudel.[27] It has even been argued that civilizations can be fixed to geographical locations such as Egypt, Mesopotamia, India, China, Turkey, Persia, Central Asia, Europe, Africa and America. Thus, it is even suggested that only relatively large empires be considered as subjects of civilizational analysis.[28]

Even if territory and geography could provide civilizations with distinctive features, it can hardly be taken as a significantly distinctive element of civilization, for civilizations in time become non-territorial. This remark in no way denies the impact of territory and geography on civilizations that, just like all social identities, take place in a territory or geography. Yet, a civilization does not only extend into remote territories but into territories with quite different conditions.

Style

Kroeber argues that style is one of the most important elements that distinguish civilizations. Style refers to the manner as against content and to form as against substance. Kroeber contends that style, primarily thought to be denoting aesthetic qualities and the fine arts, can be traced in those activities such as food (cuisine) and dress. In East Asia, for example, the food is prepared soft, and meat and vegetables are cut into bits in the kitchen, not at the table or by the eater. The food is served mixed in one dish. We see bowls and chopsticks; not plates, knives and forks. Eating habits distinguish Indian and Chinese civilizations. The fact that Vietnam eats with chopsticks, Siam and Burma without, shows that the former country lies in the orbit of the influence of the Chinese civilization, the latter two of the Indian. Similarly, the Indian subcontinent and the Middle East differ from Europe in their eating habits, as the former eats directly from its fingers while the latter uses knives and forks. Dress is, for Kroeber, another example of the distinguishing styles of civilizations. Western women's dress, dominated by a skirt and either emphasizing or de-emphasizing the anatomy of waist and legs, differs markedly from the ancient Mediterranean dress (Egyptian, Greek, Roman) in which there was no skirt as such and both waist and legs were neither emphasized nor de-emphasized. In the Far Eastern dress styles, there was no modulation of bosom, waist and pelvis and the Western skirt was replaced by long sleeved coats or draped garments hung from shoulders.[29]

Kroeber thus concludes that style is the best indicator of civilizational delimitations.[30] McNeill agrees with Kroeber: 'The only real guide historians have for assigning spatial and temporal limits to "civilizations" is a ... sense of social style.'[31] Both authors include arts and literature in style. It seems fairly evident that most civilizations develop a set of characteristic basic styles manifested in a variety of activities ranging from what are called intellectual creativities such as philosophy and literature[32] to what may be called the daily trivialities such as fashion and cuisine.

History

A common historical experience shared, or a historical process gone through, by peoples of a civilization can be said to create the constitutive and distinctive traditions, customs, institutions and characteristics. A common history could also make peoples achieve a 'civilizational consciousness' so that they identify themselves with the said civilization. A particular rule experienced throughout a period of history, or a particular threat felt for some time, or a particular movement undergone in history, could lead to the distinctive characteristics. Shared Roman rule, for example, brings Western and Islamic civilizations closer in comparison to Indian and Chinese civilizations. As noted earlier, the Islamic threat felt by Europeans has been an important ingredient of their civilizational identification. The Reformation movement, which Western Christendom has undergone, but Eastern Christendom has not, formed one of the distinctions between these two civilizations. Similarly, the fact that 'the Eastern Church remained permanently antipathetic to the idea of the Crusades'[33] may be seen as another distinctive historical experience. It can fairly be said that a common history makes a difference.

We can add to this list of elements some other elements such as customs, institutions, traditions, systems of government, military techniques, ways of production, myths, elite culture and so on. For example, for Hodgson, a cumulative tradition in the form of high culture is the determining element: a civilization is a compound culture, 'a relatively extensive grouping of interrelated cultures insofar as they have shared in the form of high culture, on the urban, literate level'. McNeill, on the other hand, defines a civilization as the totality of different and diverse groups who lived in very different ways, but were held together by 'their common subjection to rulers, whose continued dominion was much assisted by the fact that they subscribed to a set of moral rules embodied in sacred or at least semi-sacred texts'.[34] For both scholars, civilizations are composed of different and diverse groups. What unites these different groups under a common civilization is elite culture shared, for Hodgson; and common rulers they have been subjected to and symbolic and sacred texts they have been assumed to share, for McNeill.

On the basis of the analysis of those 'distinctive' elements, it can, I think confidently, be argued that civilizations are not distinguished or delimited from one another by just one criterion or a single element. They are differentiated in some degree by all of the elements outlined above. This remark should not be interpreted as a reflection of an eclectic attitude. It is very easy to dismiss it as eclecticism. Yet, civilizations are social entities that stem from the sociality of human beings and human social identification. As social identities, they all have social products. In other words, it looks quite natural for civilizations, being what men have produced and inherited, to comprise a variety of elements or human activities from language to myths. The multi-faceted character of human phenomena can hardly be explained and distinguished by one aspect or element.

A civilization is, like all social entities, simply a result of the institutionalization of human activities which form human life, and which are carried out by some people in a particular way. The differentiation of civilizations is no easy matter. Social phenomena and social identity develop as ongoing processes, implying that the composition and boundaries of civilizations could change in time. Kroeber is aware of the difficulty and submits to the authority of common sense which 'demands that we accept civilizations as units naturally given in history'.[35] Therefore, only by historical analysis, we can see what have been (and are) those civilizations with which human beings have identified (or are identifying) themselves.

Appealing to history is no absolute or final remedy due to known reasons. Not only do different historians delineate different civilizations – and different numbers of civilizations – but the same historian may present different delineations at different times. A list of civilizations by a particular historian is, just like all lists, bound to be arbitrary. Nonetheless, history is the only solution we have. Moreover, we can find some agreements on the number and types of civilizations by the community of historians. Then, it would be better to see what has been proposed.

Historically existing civilizations

We do not know who was the first author who enjoyed the idea of the plurality of civilization and distinguished separate civilizations. We do, however, know that societal differentiations have always existed. According to McNeill, the historical evidence in hand at the present time allows us to say that distinct civilizations have emerged – by diffusion from Mesopotamia – in the main regional centres of the Old World such as Mesopotamia, Egypt, Minoa and the Indus Valley between about 4000 and 1700 BC. We have archaeological evidence to show us that there had been contacts and exchanges among those civilizational centres. By AD 1000 the local civilized societies began to have inter-civilizational exchanges all over the Old World and from AD 1500 the Amerindian peoples were subjugated and incorporated into the Eurasian world system.[36]

As separate societies (and civilizations) encounter each other, men naturally see the differences between them. In the fifth century BC, for example, Herodotus knew who were the Greeks and who were not. Yet, those who were not Greeks were not, for Herodotus, the peoples of another civilization, but barbarians. The tendency to regard others as the barbarians was not, of course, something peculiar to Herodotus. It went on till the modern age and still exists, at least, in its adjectival form.

I have already said that Herder could be credited with being the first author recognizing other civilizations. While Toynbee gives the credit to de Gobineau, Sorokin appears to be of the view that credit goes to Danilevsky.[37] As always, it is probably a vain effort to look for the originator. What is certain is that, as a parallel to, in Braudel's words, the 'triumph of the plural form' of the word 'civilization' in the nineteenth century, writers began to identify various civilizations.

De Gobineau and Danilevsky supplied us with the early enumerations of the major civilizations in history. De Gobineau enumerated only ten peoples that had arisen to the level of civilization: the Indians (arose from a branch of Aryan, a white people), the Egyptians (created by an Aryan colony), the Assyrians, the Greeks, the Chinese (brought about by an Aryan colony), the Ancient Italian Peninsula (the cradle of Roman Culture), the Germanic races (who transformed the Western mind in the fifth century), the three civilizations of America, namely, the Alleghanian, the Mexican and the Peruvian. Danilevsky, in chronological order, listed civilizations or in his terms 'main culture-historical types' as follows: Egyptian, Chinese, Assyro-Babylonian–Phoenician–Chaldean or Ancient-Semitic, Hindu, Iranian, Hebrew, Greek, Roman, Neo-Semitic or Arabian, Germano-Romanic or European. To these may be added two American types, that is, Mexican and Peruvian, that perished by violent death and did not complete their life course.[38]

In the twentieth century, many authors delineated different numbers of civilizations. While some suggested fewer numbers, others gave as many as twenty-one or so. Spengler and Toynbee have been the most passionate authors claiming that only civilizations could be taken as meaningful units of historical study. For Spengler, who argues that civilizations are exclusive and impenetrable, there have been eight civilizations, or in his terms 'great cultures': Egyptian, Babylonian, Indian, Chinese, Classical or Apollonian (Graeco-Roman), Arabian or Magian, Mexican, and Western or Faustian (which emerged around AD 1000). He mentions the Russian as the next possibility. Of these nine cultures, the Mexican died by violent death, the Magian and the Russian underwent a 'pseudomorphosis'.[39]

Toynbee, who perhaps counted more civilizations than any other writer, has first identified twenty-one: the Egyptian, the Andean, the Sinic, the Minoan, the Sumeric, the Mayan, the Syriac, the Indic, the Hittite, the Hellenic, the Western, the Orthodox Christian (in Russia), the Far Eastern (in Korea and Japan), the Orthodox Christian (main body), the Far Eastern (main body), the Iranic, the Arabic, the Hindu, the Mexic, the Yucatec, the Babylonic. Second, four 'abortive civilizations' have been delineated: the Abortive Far Western Christian, Abortive Far Eastern Christian, Abortive Scandinavian and Abortive Syriac. Then, he adds five 'arrested civilizations': the Polynesian, the Eskimos, the Nomadic, the Ottoman and the Spartan.[40] Therefore Toynbee has a list of thirty civilizational units altogether.

Hodgson identifies four civilized regions of the Afro-Eurasian zone of civilization, namely Europe, the Middle East, India and the Far East of China and Japan.[41] Similarly, McNeill builds his *World History* on four major Eurasian civilizational centres from each of which a distinctive style of civilization is derived, namely those of the Middle Eastern, the Greek or European, the Indian and the Chinese.[42] Despite the dominance of the European or Western style of civilization at present, McNeill adds, others are still living. Lately, Huntington has counted seven or eight existing civilizations among which 'the clash' will be taking place in the post-cold war era: the Western, Confucian or Sinic, Japanese, Islamic, Hindu, Slavic-Orthodox, Latin American, and, possibly, African civilizations.[43] One may cite other authors with still different lists.

The reason why we find different delineations with different authors can, of course, be explained in terms of their conceptions and assumptions. Perhaps, we can find a more fundamental reason. I have earlier made the point that civilizations, being a dynamic process as are all social entities of social identification, could change in composition and boundary. The same dynamic process holds for the emergence of separate, and new, civilizations throughout history, and thus for the emergence of different lists. A civilization may be separated into two parts due to such factors as the withering of communication between the respective parts and the rise of different interpretations with regard to the identificational elements. When such a development takes place what is seen is not a commonality forming a civilization, but two (or more) commonalities forming two (or more) civilizations. For example, McNeill tells us that such a break happened between Orthodox and Latin Christendom in the early medieval centuries, and between China and Japan at about the same time.[44] I would therefore endorse the view that rejects the restricted number of civilizations, and says that there can be nothing final about the listing of civilizations in history.[45]

There can be no final list of civilizations and no final agreement among historians on the number and boundaries of civilizations. If so, then, does it mean that we fall back into the 'anything goes' anarchy of relativism or a mere intellectual sophistry? Not exactly. Even if it is not possible to make final demarcation lines between civilizations (between and of what can it be made?), it is possible, as I have already stated, to find some agreement within the community of scholars. There may be disagreements about the boundaries and compositions of the 'Western civilization', or 'European civilization', as some extend it to cover the Americas and some ex-colonial territories, or the Balkans and Russia. Yet, almost everyone agrees that there is a distinct Western or European or Christian civilization (whatever name that may be called), distinct from, say, an Islamic, or Indian or Chinese civilization. Similarly, today, the distinctiveness of Islamic, Indian, Chinese, and Japanese civilizations has been recognized by many scholars, though the nature, boundary and characteristics of each civilization could be contested. It has already been noted that there have been those arguing for the emergence of distinct Orthodox-Russian, African and Latin American civilizations.

Drawing upon the works of a variety of authors from Spengler and Toynbee to Kroeber and Bagby, Melko tells us that a good measure of consensus has been reached on the following areas in which separate civilizations have been distinguished in history:

- The Far East between 2000 BC and the present
- India between 2500 BC and the present
- Egypt between 4000 BC and 300 BC

- The Middle East between 4000 BC and the present
- The Mediterranean between 3000 BC and AD 1500
- Western Europe between AD 700 and the present
- Central America between AD 1 and AD 1600
- Western South America between AD 1 and AD 1600

Melko finds a pronounced tendency to distinguish an Islamic civilization round the Southern Mediterranean between AD 500 and the present, an Orthodox civilization in Eastern Europe at roughly the same period and a civilization in Japan, since possibly 400 BC, distinct from China.[46] Melko's comprehensive survey shows us that the degree of agreement among historians is not negligible. It is quite significant to the extent that it allows us to say historically and for certain that distinct civilizational units have existed and do exist in history, even though there is no absolute 'objective' element for delineation.

Subjective identification

I have already stated that a civilization could be defined both by common 'objective' elements and by the 'subjective' identification of people in concern. Having examined the 'objective' elements which constitute commonalities for a civilization, the question now is whether there may be found a civilizational consciousness in human beings who are said to be the members of the civilization in question. To formulate it another way, we need to see if human beings identify themselves with a social entity that we call civilization. Apparently, this is an empirical question.

However it must be stated that the distinction between the shared 'objective' elements and 'subjective' self-identification, put as a working definition, does not mean that the two processes are independent of each other. In other words, the very existence of 'objective' common elements indeed reveals that there is a self-identification of human beings with each other and consequently with all the others having the same elements or characteristics. If people have nothing in common, how could they be expected to identify with each other? The question is not, then, whether there can be a subjective identification or civilizational consciousness, but how cohesive or how strong it is. The question of degree is in turn something which depends upon the specific civilizations and historical conditions.

It has already been suggested that, since civilizations are large-scale social entities, most of them contain multiple sub-identifications in themselves, and identity consciousness in civilizations consequently tends to be less cohesive and weaker than, say, it is in the case of smaller identificational units such as states, local societies, *ethniés* and so on. It is, as already stated, a commonplace view that the existence of an external threat is a constitutive and consolidating factor of the collective identity consciousness. In terms of civilizational consciousness, it can easily be observed in the case of the struggle between the Islamic civilization and the Western civilization in history. The long rivalry and struggle between these two civilizations have been an enhancing force of common identity of the each, despite the obvious historical and theological similarities they have.

Accordingly, the existence of a high measure of homogeneity in terms of the 'objective' elements is expected to heighten the degree of cohesiveness and strength of civilizational identity. This can easily be exemplified by the Japanese civilization, which is virtually uni-national and highly homogenous, unlike the Western or Islamic civilizations, each being multi-national and highly heterogeneous. The history of Western civilization provides another exemplification. In the medieval period, when the cultural unity of Western civilization was quite high, it was possible for the Europeans to mobilize for the Crusades. Now, in the modern period, given that there is much less cultural unity than there used to be, it is very unlikely that such a mobilization of peoples could be achieved in Europe.

Perhaps one of the most typical expressions of civilizational consciousness or identity is the self-regard or self-image of each civilization. Each civilization regarded itself as the civilization (in the singular form) and others as inferior to its own or as mere barbarians. Almost every civilization from time to time considered itself to be the centre of the world and had its own, to use Gong's term, 'standard of civilization'.[47] The Greeks had it. In the words of Plato:

> When Greeks and Barbarians fight, we shall say that they are natural enemies, warring against one another, and this enmity is to be called war; but when Greeks fight with Greeks, we shall declare that naturally they are friends, and when anything of this kind occurs, Greek is sick and attacked by sedition, and this kind of enmity is to be called sedition.

And in Aristotle's:

> Right it is that Hellenes rule barbarians ... The Greeks accordingly reject the term 'slave' as applicable to Greeks, and confine it to

> barbarians ... The barbarians are more servile in character than Greeks, and are therefore prepared to tolerate despotic government.[48]

Of course it was not only the Greeks who had a sense of superiority and self-righteousness.

As Gong has shown us, the Europeans articulated a 'standard of civilization' to which the others, with whom the Europeans established contact, have been subjected.[49] Nor was it only the Europeans. Muslims saw themselves as the soldiers of the God, truly civilized, charged with bringing the infidels, certainly the uncivilized, into civilization, – the way of the God. The classical Chinese literature had it that China was the centre of the world in geographical, cultural and political terms and it was self-sufficient in all aspects without needing anything of the others.[50] A letter from the Emperor Chi'en Lung of China to King George III of Great Britain, sent in response to the latter's request for an English representative to reside in Beijing in order to oversee mutual relations, boldly expressed it thus: 'Our celestial Empire possesses all things in prolific abundance and lacks no products within our borders. There was, therefore, no need to import the manufactures of outside barbarians in exchange for our produce.'[51]

In sum, every civilization had/has its own image as an expression of its identity, derived from its common elements and its encounters and interactions with the other societies, and very often articulated in opposition to the external 'barbarians'. However, it must here be stressed that the self-image of a civilization being mostly defined in opposition to the others in no way implies that civilizations are in a constant conflictual state. For example, Toynbee notes that in the thirteenth century, in Uiguria, Nestorianism and Buddhism, two universal religions, were actually living cheek by jowl in a *modus vivendi* of mutual toleration.[52]

Conclusion

No matter that the distinctiveness of the elements and the consciousness of subjective identification or internal cohesion is questioned, and despite the existence of disagreements on the delineation: civilizations do exist. One is certainly justified to ask how meaningful or 'intelligible' civilizations are as units in historical and social studies. They are intelligible units of analysis like the other more familiar or immediate units such as nation-states. It is not because civilizations are more inclusive and thus relatively self-explanatory, but just because man is a social and creative being to establish relations and identifications of different types at different levels. Civilizations are not the only identification of men as there are others – for example, nation-states, *ethniés*, localities and various kinds of groups. How effective or directing then is a civilization upon the behaviours and actions of human beings? To what extent is individual identity determined by civilizational identity?

It could be asserted that the impact of civilizational identity upon the identity and actions of the individual is not institutionalized to the degree of being exclusive as it is in the case of political or organized social identifications. If you do not have identification with the Chinese civilization, it does not prevent you from having your meal by using chopsticks. However, if you do not have identification with the People's Republic of China, you cannot bear a Chinese passport. Even if you do not have the identity of Western civilization, you may dress and eat like a European. Yet, you may not possess a right to permanent residence and work in the United Kingdom, if you are not of British nationality.

Nevertheless, civilizations do influence individual or collective actions. First, civilizations being social entities affect the socialization of individuals and individuals are partly moulded and shaped by civilizations. It is true that it is not forbidden to use chopsticks, but it is also true that not having been raised in the Chinese civilization makes it difficult for you to use chopsticks. Second, individuals take into consideration their civilizational identity like all other collective identifications when they act according to specific cases. We can but say that the influence of civilizational identity on the actions of human beings is less than that of smaller, organized and more immediate collective identities.

Civilizations and universal history

To recall the final question of the third chapter, how could civilizations form a basis for a historical conception of universal history? If civilizations are distinguished from what is called simple, hunting-gathering societies, one thing we could confidently say is that they have been persistent and self-transforming throughout history, as we know it. Moreover, we can also say that civilizations have come to prevail all over the world, whether they had begun separately in different locations or emerged through borrowing from each other and, in the final case, emanating from one centre which is said to be Mesopotamia. It has been said that all simple societies have now been brought under civilized forms of societies or administrations with the development of mechanical means of transport and communication.[53] Such an account of world history is not an idealistic or universalistic one as it simply takes civilizations to be the kind of societies with organized institutions and specialization.

One can argue that civilizations when conceived as distinct societal entities can be considered as a basis for a historical conception of universal/world history. If world history is defined as history 'taking the world as a whole', of all societal entities/identities known in history, it is via civilizations that we get the nearest to 'taking the world as a whole'. The reason is that civilizations, being massive, large-scale and long-lived, could provide us with a picture of the world (if such a picture is ever possible) in its spatial and temporal extensions known to people at a particular point in time. Spatially, because, *pace* Spengler, there have always been encounters and relations between civilizations. Temporally, because civilizations have outlived all other known societal entities and they enable us to establish a historical continuity between societies of different ages through affinities and affiliations. The affiliation of the modern European societies with the Greek and Roman societies is self-evident.

Seen as such, interactions between civilizations have not always been even. Sometimes, one or more civilization(s) has (have) predominated over others in terms of their power, organization, vitality, cohesiveness, potency and so forth, and of being the main attraction centre of the world. It was the Middle Eastern civilizations in the third and second millennia BC, later it was Graeco-Roman, then Islam and now it has been the European/Western civilization. The Western civilization has far outstripped the others so that it has a global or world-wide impact in the contemporary period. Other civilizations such as the Islamic, Indian, Chinese and Japanese having been subjugated, McNeill speaks of 'the establishment for the first time in world history of a genuinely global cosmopolitan civilization, centred upon the West but embracing all the other varieties and cultural traditions of mankind as well'. He adds that interaction between the heirs of the Western, Islamic, Hindu, Chinese and Japanese civilizations has been and in the foreseeable future promises to remain, 'a central axis, and perhaps *the* central axis', of world history.[54] It is upon inter-civilizational interactions that a world history can adequately be built.

Civilizations and international systems

Civilizations are not, as I have already emphasized, integrated social entities as are some other social identifications (for instance, states, local communities, groups ... and so on). They have been not only multi-dimensional but also multi-organizational. Most civilizations have been, most of the time, multinational and multi-organizational in terms of body politic. They have diversity and conflict within and their boundaries are imprecise.[55] In other words, most civilizations have a kind of, to use a term familiar to students of International Relations, 'states system' or 'international system'.

International systems are simply defined as a grouping of, or formula for, the multiple socio-political units in order to have smooth interactions with each other. They set frameworks for the peaceful or smooth coexistence of multiple units. As is well known to the students of International Relations, an international system is composed of multiple units (nation-states) whose behaviours become a consideration for each other when they form their policies vis-à-vis each other. International systems may be 'anarchical' in the sense that the members of the system do not acknowledge a common supreme power above themselves and they interact within the framework of a loose association. International systems may be imperial or 'hierarchical' in the sense that some form of a centre is recognized or emerges to lay down the rules for the interaction of multiple units and the supremacy of the centre is acknowledged, at least nominally. Then, the international system may be taken as a response to the question of the peaceful coexistence of multiple units.[56] Yet, in time, it becomes a unit comprising multiple socio-political groupings of human beings just as much as civilizations.

There is a close similarity between international systems and civilizations. I have already stated that most civilizations have, for most of the time, contained, or given way to, an international system. Examples of the international system defined by the students of International Relations show the link between civilizations and international systems. Wight, for instance, identifies three systems of states: the Hellenic-Hellenistic or Greaco-Roman, the Chinese between the collapse of the Chou Empire in 771 BC and the establishment of the Ts'in Empire in AD 221, and the Western systems.[57] Bull gives five examples of 'international society' as follows: the classical Greek city-state system, the international system formed by the Hellenistic kingdoms in the period between the disintegration of Alexander's empire and the Roman conquest, the international system of China during the period of Warring States, the states-system of ancient India, and the modern states-system which arose in Europe and is now worldwide.[58] We see that Wight and Bull's examples of states-systems are associated with civilizations. Each took place in, or started from, a civilization.

Melko enumerates the notable state systems as follows: the Modern European, the Italian Renaissance, the Hellenistic, the Greek City-States, the Ch'un Ch'in-contending states period in China, the Han and T'ang Empires of China, India and South East Asia throughout most of their history, Islamic history generally, modern Latin America.[59] 'Seven major international systems which preceded the one with which we are familiar in the twentieth century' are cited by Northedge. Namely, the ancient empires of the Near and Middle East (exemplified by the Egyptian, the Assyrian and the Persian empires), the Chinese empire, the ancient Indian empires before the Mogul empire, the Greek city-states system, the Roman empire, the Byzantine empire, and the Medieval system from which the modern international system directly or indirectly sprang.[60] Recently, A. Watson has identified 'the more important and well-documented' states systems of the pre-modern period as the Sumerian, the Assyrian, the Persian, the Greek, the Macedonian, the Roman imperial, the Byzantine, the Islamic, the ancient Indian, and the Chinese systems. To these he adds the medieval and Modern European international societies and the global international society.[61] Similarly, as seen, the majority of Melko, Northedge and Watson's lists correspond to the typology of civilizations by most historians. Those international systems have largely been intra-civilizationally international systems.

Of course, an international system does not necessarily remain confined to the boundaries of one civilization. This is partly because the delineation of the boundaries of a civilization is much more difficult to sustain than are those of socio-political groupings which constitute an international system. Also, an international system could take place between, not only within, civilizations due to civilizational encounters, exchanges and relations. In other words, an international system can be multi-civilizational. For example, Wight gives three examples of inter- or multi-civilizationally international systems, in his terms 'secondary states-systems': the Roman-Persian system, the Near Eastern system in the latter half of the second millennium BC, and the Mediterranean in the twelfth and thirteenth centuries AD.[62]

Upon the basis of the historical record, one is justified in arguing for another and more embracing multi-civilizational system, namely 'the Afro-Eurasian international system', to which I shall return later in Chapter 7. It can, I think, be fairly proposed that an international system which is uni-civilizational is likely to have a higher degree of systemness than the one which is multi-civilizational. A comparison of, for example, the Greek City-States system with the Roman-Persian system suffices to show this.

If it is possible for an international system to embrace more than one civilization, then, it means that there may be an international system comprising all civilizations. In other words, when all civilizations have sufficient relations with each other or are linked to each other in sufficient degree, a worldwide international system or global system could emerge. The emergence of a global international system means that societies with distinct civilizational identities have come into contact. The increasing exchanges between societies with distinct civilizational identities may lead to the loosening of the very civilizational identities. It has been widely agreed that such a process, the development of a global international system and loosening of civilizational ties, happened with the modern international system which emerged in Europe from the sixteenth century onwards and became worldwide or global in the twentieth century. The modern international system is then a multi-civilizational system with both inter-civilizational and trans-civilizational processes. The remaining chapters examine the modern international system.

6
Modern International System I or No Rock without a Flag

What do I mean by the 'modern international system'? As already stated in the previous chapter, it refers to the international system that originally emerged in Europe in the sixteenth century or so and spread worldwide through a process of expansion and globalization by the twentieth century.[1] In other words, the modern international system embraces the whole world. It is an inter-civilizationally international system, including not only more than one civilization, but also all civilizations in the world as we know it at present. That the modern international system embraces the whole world means that there have occurred enough global interactions among the communities or societies all over the world, for people living in a particular community to be influenced by the events taking place in other communities of the world. Or, it could mean that there are some processes or characteristics which are inter-socially shared in the world or which are globally prevalent. The modern international system is thus not just inter-civilizational, but trans-civilizational.

A worldwide system

That we live in an economically interdependent world is obvious. That the result of American presidential elections has immediate effects upon the balance of payments of, say, Turkey via possible fluctuations in the value of the dollar or via the foreign loans policy of the new administration; that the decision taken by the European Central Bank makes its impact on the London stock exchanges and financial markets; that the new investments of the Japanese firms could make some American workers go unemployed; that the policy of the OPEC countries may significantly influence the budgets of European car drivers; that French agricultural policy or their import regime could change the crops of the African farmers; that the Asian crisis of the 1990s influenced the corn farmers in the State of Iowa; and that the mid-February crisis in the Turkish economy in 2001 led to falls in the stock markets of Argentina: all these facts are well known.

It is also well known that there have always been interdependencies between the trading partners, not only in the modern international system and at this age. Yet, it could also be said that the extensive and intricate network of economic exchanges with which we are familiar now were not the case all over the world two hundred years ago. It can fairly be argued that such a high degree of interactions among societies in economical terms is something unique to the modern international system. The extent and significance of economic interdependence has been acknowledged by the students of international politics as well. For instance, a textbook of international politics, first published in 1987, appeared with a new chapter on 'the World Economy' in its second edition.[2]

Sovereign nation-states

That the modern international system covers the whole world can be observed in the principal form of legal-political organization, that is to say, the centralized sovereign state or nation-state. Today, the sovereign nation-state has been adopted by all societies as a form of political organization. Even the Papacy has taken the form of a sovereign state, that is, the Vatican. The prevalence and pervasiveness of the sovereign nation-state in the contemporary period is well described by the French poet and thinker Valéry:

> Every habitable part of the earth, in our time, has been discovered, surveyed and divided up among nations ... There is no rock that does not bear a flag: there are no more blanks on the map; no region out of the reach of customs officials and the law; no tribe whose affairs do not fill some dossier and thus, under the evil spell of the written word; become the business of various well meaning bureaucrats in their distant offices.[3]

The modern international system is, then, a system of sovereign nation-states. The states are sovereign because they claim to be the final and absolute authority in their respective realms and they deny the existence of any other authority. They are sovereign because they hold a monopoly of 'making and unmaking law' and they claim to be the only centre having the monopoly of the legitimate use of physical force within a given territory. These features of sovereignty that modern states are said to have assumed have been formidably expressed by the modern scholars and come to be accepted by the students of mainstream International Relations. Bodin, in his *Six Books of the Commonwealth,* argued that the power of 'making and unmaking law' was the quality that comprised all the rest of the attributes of sovereignty. Weber forcibly defined the state as a political community that successfully claims the monopoly of the legitimate use of physical force. Sovereignty is, according to Hinsley, 'the idea that there is a final and absolute authority in the political community and no final and absolute authority exists elsewhere'.[4]

However, the state's claim to sovereignty can always be contested and it could also be stated that it is not the only quality of the state. As is well known, in *Manifesto of the Communist Party,* Marx and Engels declared the modern state as 'a committee for managing the common affairs of the whole bourgeoisie'.[5] Sovereignty hence lies in the bourgeoisie! Ironic as this sentence is, it implies the fact that the state's claim to absolute sovereignty and to the monopoly of the use of physical force has hardly been realized in practice. As Hoffman vividly observed, the state has always shared power with some other groups.[6] However, no matter how frequently the state's claim to sovereignty is challenged and no matter how complex the state and theorizations about it are, the claim to sovereignty and some idea of it remain persistent.[7]

The modern international system is not only a system of centralized, separable, sovereign states. It is a system of nation-states, meaning that states were nationalized or that states created nations. In other words, the modern state is not just a collection of public institutions differentiated from other social institutions. A state either embodies a nation or designs it. It is the principal focus of collective-social identifications. National identity has largely been defined on the basis of the citizenship of a state. Most scholars cite the existence of a form of body politics among the elements of nation. For Hume, stating in his essay *'of national characters'*, a nation was simply 'a collection of individuals who, by constant intercourse, come to acquire some traits in common'. Diderot and D'Alembert, defining it in *Encyclopédié,* referred to a people 'obeying the same government'. A nation is, for Sieyés, 'a body of associates living under one common law and represented by the same legislature'.[8] While the peoples of a nation 'desire to be under the same government, and desire that it should be government by themselves or a portion of themselves exclusively', according to Mill's view, Renan states that a nation has 'a common will in the present'.[9]

The close association between state and nation can be derived from etymology as well. According to the *Oxford English Dictionary*, the first appearances of the modern meanings of the words took place around the same period. The state as 'a particular form of polity or government' appeared in 1538. The word nation, as 'an extensive aggregate, so closely associated with each other by common descent, language or history, as to form a distinct race or people, usually organized as a separate political state and occupying a definite territory', was cited from 1300 onwards.

Historically, either nations, which were already there before the emergence of the modern state, came under the jurisdiction of the state (as was the case with the English and French nations) or the emerging state created a nation (as was the case with most other nations). Even in the first case, Smith rightly observed, the elements of design were available in the formation of national identity, that is to say, the Jacobin nationalism and Tudor and Stuart centralization.[10] The state and nation are linked to each other. The nation, according to Smith, 'signifies a cultural and political band, uniting in *a single political community* all who share an historic culture and homeland'.[11] It would not be wrong if one asserts that the concept of nation is understood as signifying a collectivity of human beings who have political rights and obligations. Even the smallest states claim to form a nation. States do not only use legitimate physical force upon human beings but also create some idea of fellow feeling among those who are under their jurisdiction, hence the creation of national identity. Today, we see an Iraqi national identity distinct from the Syrians in spite of the existence of so many commonalities between these two peoples such as original descent, religion, language and so on.

The modern international system is thus composed of sovereign nation-states. Furthermore, sovereign nation-states remain as the principal units of the modern international system. The nation-state is still the main agent in international relations in spite of much talk about its demise. Even those who argue for the rise of multinational corporations concede this. John Herz, for instance, initiated considerable debate by making, then retracting, predictions about the coming demise of the territorial state in the face of powerful innovations in military technology. Keohane and Nye retreated from their position in the early 1970s.[12]

Moreover, the movements which question the legitimacy of the nation-state such as communism and Islamic fundamentalism not only located their community of 'proletariat' or 'believers' within the bounds of nation-states but also seem to have got on well with the established practices of nation-states. The nation-state, though a Western creation, has been adopted by other cultures and civilizations. Piscatori, for example, argues that most Muslims have come to terms with the idea of the nation-state and most have come to accept the concepts undergirding modern international law. His conclusion is that Islam, at present, chimes relatively easily with the prevailing international order.[13]

Political theorists and historians continue to draw conclusions about the enduring ability of sovereign nation-states. Two scholars, one political scientist and one historian, express the persistence of the nation-state as follows: 'Even at its most ideologically pretentious', writes Dunn, 'the species has not yet conceived a practical form in which to transcend the nation-state'. And Kennedy declares that 'the international system, whether it is dominated for a time by Six Great Powers or only two, remains anarchical – that is, there is no greater authority than the sovereign, egoistical nation-state'.[14] Of course this does not imply an ahistorical continuity of the nation-state, as Morgenthau, the uncompromising realist, stated: 'Nothing in the realist position militates against the assumption that the present division of the political world into nation-states will be replaced by larger units of a quite different character.'[15] Yet, it shows that the modern international system is globally prevalent in its political structures.

Global awareness

The modern international system is worldwide not only in economic interdependencies and global prevalence of the nation-state as a form of political organization. It is worldwide also due to the existence of a global awareness among the individuals and societies of the world. That local peoples living in a particular zone of the Earth could have the knowledge of peoples, events and happenings in other zones looks quite natural to us in this age of speedy communication and transportation. People know that there are other fellow human beings in far-away societies not just as a result of mere saying or guessing but because they are aware of them with some identifiable quality. The Turks knew that there were 'franks' living in 'frankistan' (the land of the infidels) centuries ago, but now they know that those 'franks' called the Dutch live in the Northwestern part of Europe, and that they are Protestants who have some distinctions from those franks called the 'Italians' in the Southern part.

Moreover, the sense of awareness of people goes beyond the knowledge of others. People think and know that happenings and events of other places could and do influence their daily life. When the headline of their local paper reads 'Iraq enters Kuwait, Bus fare goes up', the people of a Midlands town in England do not seem astonished and do consider that connection between the two events is possible. Similarly, the Turkish commentators see a link between the Oklahoma bombing incident in the United States in April 1995 and a decrease in the international appeal of the Kurdish Workers Party (PKK) on the grounds that the horrendous event will make the US policy-makers and public opinion withdraw their assumed support for the PKK. There is no need to extend the list of such examples. It can justifiably be asserted that some kind of global consciousness exists: a consciousness that we all live together and are interdependent upon each other. The modern international system is therefore worldwide both in its 'objective' elements as outlined in economic and political realms and in its 'subjective' element of some kind of global consciousness.

Defining characteristics

That the modern international system is global and worldwide and that this is unique to it do not suffice to define and understand it. We need to look at it more closely. So, now I would like to examine the principal and defining characteristics of the modern international system.

Decentralization

The first striking characteristic of the modern international system is that it is a decentralized system and it is composed of separable units. In other words, the quality of systemness mainly comes from the interactions of differentiated units, not from an effective embracing body such as an imperial system or an inclusive idea such as religion. That is to say, it is not a centralized system. The units of the system, though in interaction and having some commonalities, preserve their individualities and distinct identities. To use the familiar expressions of the students of International Relations, the modern international system is an anarchical system, having no central and common authority. However, this characteristic, anarchical organization, is not something unique to the modern international system.

Logically, the coexistence of multiple units may take place under three frameworks. In the first case, we see an isolated situation in which different units are separated from each other, having no or negligible relations with each other. Each unit exists by itself. In this case, we cannot, of course, speak of any inter-societal or international relations and thus no international system.

Second, multiple units may exist under a single centre of authority that embraces or purports to embrace within a common framework the different societies of the known or reached world. Such a framework could be embodied in an 'empire' or a world state. One of the units is explicitly or implicitly recognized by others as superior, or it exerts a supremacy over all the others. Here, we see inter-societal relations and we could rightly speak of the existence of an international system. Such a system is, since it has a common centralized authority, called 'hierarchical' or imperial in the terminology of the students of International Relations.

The third framework refers to a situation in which there may exist a loose association of independent communities or societies, each unit enjoying some degree of self-government. None of them is recognized as having, or being capable of imposing, an authority over the others. There is an association among them, coming from common interests or characteristics or circumstances shared and thus being created in interaction with each other. Here, too, we can speak of the existence of an international system. In sum, two types or patterns of international system could be defined according to position of the units in relation to each other. The first pattern may be described as centralized, hierarchical, hegemonical or empire, world-state or suzerain systems. The second one might be named as decentralized, anarchical, independent and so on.[16]

The modern international system is a second type of international system and there have been others as well. The classical example is the Greek city-states system. Scholars add Chinese Warring-States system, the Indian system before it was centralized by the Mogul empire and Hellenistic kingdoms before they were conquered by the Roman Empire. Most of the pre-modern international systems fall into the first pattern, that is, hierarchical or imperial system. Examples could be cited as the Near Eastern empires, the Persian Empire, Roman Empire, Chinese Empire, the Medieval European System ... and so forth. The organizational form of the modern international system is not unique to it, as there have been other anarchical systems. Then, how this could be a defining and, more importantly, distinguishing characteristic is the logical question. True, this character is not unique and distinctive. However, it is still defining and in one sense distinguishing as the core system, that is, the medieval system, from which the modern international system sprang, was a centralized system. The anarchical character of the modern international system is therefore distinctive of it from its original core system, not from the international systems that existed in history.

Nationality

The second characteristic of the modern international system is that it is, as already noted, a nation-state system. The system is not just composed of separable units, but the units, or majority of them, are of a particular type, that is, nation-state. The basic and prevalent form of collective-social identification in the modern international system has been nationality. This led to the weakening of the impact of both sub- and supra-identifications of human beings, such as local groups and civilizations. Expressed in another way, the system is based upon a particularistic identity, rather than a universalistic identity, leading to a plurality of units. Yet, with the location of this particularistic identity within the boundaries of the nation-state, the plurality has been limited.

The move away from universalistic identifications perhaps accounts for why it is the modern international system that has become trans-civilizational and global. Most scholars agree that this characteristic of the system, of units being nation-states, is distinctive. The state and nation, and thus nation-state emerged in the modern period, as we know them now.[17] Nations and nation-states which firstly emerged in Western Europe from the fourteenth century onwards spread, initially, to European colonies and then all over the world in the second half of the twentieth century. Some of the techniques and features that we now associate with the modern nation-state were known before. McNeill, for example, tells us that long before the rise of modern bureaucracy 'a political order dominated by bureaucrats, i.e. by administrators whose power inhered in their offices rather than in their persons, had been well developed under Hammurabi' around 1700 BC. He also tells us that it was the Persians and Assyrians who first used professional, standing armies.[18] Nevertheless, given its extensiveness and the dimension of national identity, the modern nation-state as we know it (though some of its techniques were known long before the modern period) is then a distinctive characteristic of the modern international system.

Humanism

The third characteristic of the modern international system could be suggested as the change in values and the point of reference in social life of peoples from the divine to the humane. A change which was prompted by the Renaissance and the Reformation in Western Europe from the fifteenth century onwards was later accelerated by the scientific and technological revolutions in the seventeenth and eighteenth centuries. With this transformation, ideas such as secularism, humanism, rationalism, individualism and democracy have become prevalent. Although it is possible to trace the secular origins of the modern international system back to, surprisingly, the Augustinian separation of the City of God from the City of Man, or to the address of Pope Gelasius I to the Emperor Anastasius at the end of the fifth century, in which he stated that 'this world is ruled by two things, the sacred authority of the priesthood and the kingly power',[19] secularism emerged in the modern period together with other developments such as humanism and rationalism.

While Burckhardt sees the origins of humanism back in the beginning of the fourteenth century, Oakeshott considers rationalism as 'the most remarkable intellectual fashion of post-Renaissance Europe'.[20] Similarly, it has been suggested that individuality as opposed to collective identifications such as race emerged in the modern period and one of the principal features of the modern international system, especially in its European core, has been individuation.[21] Democracy as the integration and participation of masses in the government is said to be another prevalent feature of the modern international system. In this sense, it emerged in relation with the other features outlined above and gained strength after the seventeenth century.

These features have been closely associated with nationality and the nation-state. The state became national only when the people became a nation through their participation in the affairs of the state, something that happened with democratization from the seventeenth century onwards. Without the idea of the liberty and equality of each individual it was not possible for the state to become national. Accordingly, without a belief in the value and ability of individuals it was not possible to conceive the participation of individuals in nation and the state. Moreover, to make human beings participate in the affairs of the state, the effective and frequent intervention of the divine or those associated with the divine needed to be eliminated and the affairs of the state and nation had to be made human or transferred from Heaven to Earth.

Though linked with nationality or nationalism, rationalism, humanism, secularism and democracy in their essence and scope had universal messages and thus created an association with some idea of universalism. It could perhaps be said that this universal appeal, not restrictive and exclusive like other universalistic formulations such as

Christendom or Dar-al-Islam, contributed towards the modern international system becoming worldwide. It is certainly easier to get people united around the idea of mankind, of which everyone is a member, rather than of God, in whom not everyone believes (or not in the same way); and around this world, in which everyone lives, rather than the other world, which no one is certain to reach. Lerner was right in saying that 'the modern international system was, by nature of its modernity, born at the moment God died'.[22]

The prevalence of these features as the characteristic of the modern international system does not mean that there had not been any trace of them before as one could find some of them in the ancient world, especially in Greece. What is characteristic of the modern international system is that these features have been pervasive and prevalent in a greater degree and with a worldwide appeal and impact which was never the case in ancient Greece, for instance. It is true that those values have not been universally adopted by the societies of the world. However, it is also true that the appeal of the challengers has been much more limited.

Industrialism

A final characteristic of the modern international system can be given as industrialism, meaning that industrial techniques and production are the dominant mode, or the significance of agricultural or land-based production and techniques has decreased sharply. Defined simply, industry or industrial production refers to an excessive use of machine work or mechanization, not just the use of simple tools as in the case of agriculture, and to the process of production in which the end product is not just a result of the labour of one worker (thus creating a chain of production) as is the case with agriculture. Some scholars indeed examine the modern international system in association with the rise of capitalism and industrialism.[23] The impact of the Industrial Revolution upon the modern international system is well-known. One is justified in saying that this characteristic is unique to the system and one of the most important factors in its becoming global.

I have earlier made the point that to understand the modern international system we need to see its principal characteristics. However, an analysis of the principal characteristics of the system is not sufficient to understand it. This is because in order to understand or know something we need to know how that thing has come to be what it is. I will now turn to the history of the modern international system, to an examination of how it emerged and developed.

Emergence of the system

To analyse the historical development of the modern international system means to scrutinize its principal characteristics, to examine when and how they appeared. There is a common agreement that the principal characteristics and features of the modern international system originally emerged in Europe. The system has become worldwide and global with the exportation of those characteristics to, and/or adoption by, the non-European societies/systems in the course of the European expansion into, and/or European interactions with, the non-European world.[24] If this is so, then the development of the system may be examined under two periods or headings: first, the emergence and development of the European international system and, second, the rise of a worldwide international system.

It must here be stated that those two processes of the emergence of the European system and the rise of a global system cannot easily be separated from each other as the emergence of the modern European international system is said to be closely related to the incorporation of other societies or systems into the European system. In other words, the emergence of the European international system went hand in hand with the European expansion into other areas. Then, it could be argued that there were two aspects of the rise of the modern European system: on the one hand we see a self-transformation of Europe through which Europe acquired the principal characteristics with which we are familiar, and on the other, the process of European expansion by which the system became a global one. Accordingly, the self-transformation of Europe and its expansion are not independent of each other. Nevertheless, for the sake of simplicity, I shall examine the two processes in turn.

Europe's self-transformation

By European self-transformation, is simply meant transition from the Medieval to the Modern Age. What happened after the transformation has already been summed up: separate, centralized and sovereign states which later came to be called nation-states, secular politics, the rationalization of human thought and of human relations, the growth in science and technology, the Enlightenment and the Industrial Revolution. To understand the transformation, it is necessary to have some idea about the pre-transformation era. How did Europe look before the transformation?

Medieval background and unity

Described briefly, the medieval system was, it has widely been agreed, characterized by unity. The unity was formed through the sense of being part of the united Christendom and also by the existence of authority centres claiming universal jurisdiction all over Christendom, in fact, all over the world. The Pope and the Emperor are said to have claimed universal authority in their respective jurisdictional areas and, frequently, competed with each other. Despite the existence of an innumerable multitude of governmental units in reality, the idea of the unity of the world in all respects was prevalent and it was believed that the world must be unified in politics as well. Continuity and stability in society, it is so argued, grew on this unity. This medieval universalism constituted the basis of all social relations that were basically carried out under the framework of religious creeds. Emphasis was on the concept of right or goodness rather than interest or the requirements of the particular case. Such is the picture of Medieval Europe.[25] According to this picture, then, it looks clear that the 'Medieval European international system' falls within the second pattern of inter-societal coexistence, or within the first type of international systems (empire type) I have described in this book.

The presentation of the medieval system as such is by no means accepted by all scholars. In questioning the idea of medieval unity, Barraclough makes the point that the said unity did not go beyond being a theory. It was not a factual situation and was never expressed in political terms, and it never took the shape of a single organizational unit comprising the whole of Christendom, not even Latin or Western Christendom. The idea of a united medieval empire was not anywhere near the truth. Indeed, he argues that the notions of nation-states and balance of power which were commonly associated with the modern period held sway in the Middle Ages. 'As soon as Europe had recovered from the anarchy which assailed it in the ninth century, the elaboration of a system of sovereign states and the creation of a European balance of power began ... the European balance of power, as we know it; the articulated, interlocking system which has dominated international relations without intermission down to to-day, was operative from the start of the twelfth century.'[26] The forceful remarks of Barraclough, no doubt, have some truth. It is true that there was no integrated, centralized Christendom or Western Christendom, and, it is true that the concept of unity was more of a theory than a fact. However, nobody depicting the medieval unity as such argued for the contrary. Yet, many, I suspect, deny Barraclough's view that the balance of power, as we know it, was operative from the start of the twelfth century.

The medieval unity cannot, I think, be underestimated by simply characterizing it as a mere theoretical notion or claim. The claims of universal jurisdiction did not result in a single political organization, yet they were effective, that is to say, they influenced the actual behaviour of men. Without the existence of some degree of unity, how else was it possible to organize the Crusades, involving a Europe-wide mobilization?

Furthermore, in the medieval period, as Southern observed, 'areas of authority shaded into each other and overlaid each other with little relation either to geography or history. No political boundaries survived in their entirety the death of a ruler; they were all subject to chances of domestic change, marriage, dowry, partition, death and forfeiture.'[27] Yet, we know that the centres of authority claiming universal jurisdiction were not shaded and they survived the death of rulers. The fact that there were no settled boundaries and overlapping jurisdictions could count for the idea of medieval unity. The medieval system, then, involved an effective universalism, represented and exercised by the Church which is said to be 'the real state of the Middle Ages in the modern sense'[28] and an excessive particularism, characterized by the multitude of the units of authority. It has been argued that this fact, universalism on the one hand and particularism on the other, hindered the development of sovereign, centralized nation-states.[29]

The universalism of the Middle Ages continued to be effective, in varying degrees, well into the early modern centuries. The fact that the term *Respublica Christiana* was referred to in the Treaties of Utrecht in 1713 and the title 'Holy Roman Empire' was abolished only in 1806 could be taken as an indication of how the idea of medieval unity was effective. The existence of the Ottoman threat is said to be one of the factors in the continuation of medieval universalism. We may wonder about the coincidence that the defeat of the Turks and the last usage of the *Respublica Christiana* took place in the same quarter: the first happened in 1699 with the Treaty of Carlowitz, and the second, just fourteen years later. According to Kohn, the expulsion of the Turks from Central Europe and the extinction of the Spanish Habsburgs, the last dreamers of a Christian world empire, at the end of the seventeenth century marked the definite end of medieval universalism.[30]

Consolidation and disintegration

Europe transformed from this medieval universalism to a nation-states system in the modern age. In contrast to the overlapping jurisdictions that prevailed in the medieval system, a series of well-consolidated states emerged in the sixteenth and seventeenth centuries. In other words, centralized states, denying both universalism and particularism of the medieval period, came to control the political and social landscape of Europe. The states of modern Europe were, according to Oakeshott, the outcome of movements of consolidation and disintegration. By the movements of consolidation, local independencies were destroyed. By the movements of disintegration, medieval universalism broke-up.[31] The outcome of these two processes was the emergence of centralized states.

Medieval particularism, the existence of an innumerable multitude of governmental units, was perhaps the inherent potentiality of the medieval system to disintegrate. From the thirteenth century onwards, units other than the two universal centres of authority began to get recognition and the united Latin Christendom is said to have had its last meeting at the first quarter of the fifteenth century, at the Council of Constance of 1416. Innocent III proclaimed the independence of the king of France, who 'recognizes no superior in temporal affairs', in 1202. According to Wight, the Concilar Movement, which started at the Council of Constance, when it ended in 1449, 'left a strengthened papacy on a narrower foundation, and an international anarchy of strengthened secular powers'.[32] The great divide is said to have occurred between the Catholic states and the Protestant states with the Reformation. By the mid-seventeenth century, not only the Catholic and Protestant states acknowledged each other, but also the competing nations confirmed the independence of each other, and the secular character of politics became established, as symbolized by the Treaty of Westphalia. The Renaissance and Reformation enhanced the movements of disintegration through challenging the traditional certainties of medieval society.[33] Without the break-up of the medieval universalism, it was not possible for the separate states to come into existence.

Equally important was the integration of innumerable governmental units within certain centres of authority which we call states. Of course, centralized states were not just formed through the coming together of various units. The process of political consolidation was undertaken by some persons who were already recognized as rulers with authority of some sort. To this were added the socio-economic and cultural conditions. The process of the consolidation of states in Western Europe may be examined from four points of view.

First, the application north of the Alps of the techniques of government which had first been worked out in Italian city-states. For instance, the merging of noble and burgher classes into a single body politic, and the use of standing professional armies. These techniques were first adapted by the French and English kings.[34]

Second, the Renaissance and Reformation not only led to the disintegration of medieval universalism but also contributed to political consolidation as a national consciousness of some sort and the state embodying it, and providing some degree of stability and unity became more and more desirable in the following turmoil. Moreover, in Protestant countries, governments took over most of the properties that were formerly owned by the churchmen and acquired the right to appoint, or at least to approve the appointment of, the leading clergymen. Watson argues that the Reformation, especially the movement associated with Calvin, endorsed the emerging separate states rather than the universal Christendom based on the Roman Empire. Through a return to the Bible, men saw the independent Jewish states of the prophets.[35]

Third, the rise in military technology and the sophistication of weapons made it prohibitively expensive for local rulers and impossibly complicated for universal authorities to exercise organized violence. New military technology, together with standing professional armies, made it easy for kings or national rulers to defeat and conquer local rulers. The same technology was not, however, convenient for universal conquest. For instance, heavy artillery and the supply of powder and other raw materials made it difficult for rulers to transfer them in long-distance expeditions, thus making it difficult to achieve a universal dominance.

Finally, we should add the role of the development of market economy and the expansion of Europe or the Discoveries in the consolidation of state structures. Market economy made the producers and merchants of different localities interdependent. The traders aligned with the central rulers (kings) against the exclusive practices of the local rulers. With this alliance, the states gained the capacity to finance long and expensive expeditions. The European expansion and the Discoveries were mainly financed by, and made on behalf of, the kingly powers. The flow of wealth from colonies in turn strengthened kingly or state power.

Europe as a system

By the mid-seventeenth century, the movements of consolidation and disintegration were practically completed. Sovereignty and independence of the states were recognized, and differences were acknowledged.

The developments in Europe from 1500 onwards such as the Discoveries, Reformation, Renaissance and the growth of science, according to McNeill, resulted not in discovering and enforcing a universal truth, but in cultural and intellectual pluralism. In other words, 'Europeans discovered that they could agree to disagree'.[36]

With the new era, moving away from medieval notions of right and goodness, rulers came to behave, and agreed in acting, according to the idea of the interest or the reason of state, and the exigencies of the particular case. Although the political identity of Europe came to be defined as a multiplicity of states by the end of the seventeenth century and every state asserted its sovereignty and independence, a sense of unity and some kind of common identity of Europe prevailed. The notion of Europe as a system constituted by its elements came to be established.

Up until the eighteenth century, the notion of Europe had been understood either in purely geographical terms or as being interchangeable with the Christendom. It gained its political and cultural meaning in the mid-eighteenth century. Though Voltaire considered 'Christian Europe' as a 'great republic', he stressed its diversity and 'the same principles of public and political law'. From the mid-eighteenth century, writers, for example, Burke, began to call the international system that emerged in Europe 'European' rather than 'Christian'. Similarly, from the end of the eighteenth century onwards, writers of international law began to use 'European' law of nations, not just a general reference to a 'law of nations'. This can rightly be taken as an indication of the rise of some sort of self-consciousness among the Europeans.[37]

All this shows that by the eighteenth century the emerging states in Europe came to constitute a system, not just a system of interdependence based upon exigencies or structural positioning, but also a system that grew upon a common identity of Europeanness and some degree of self-consciousness.

European expansion

The second aspect of the emergence and the rise of the modern European system was, as I have already said, the process of European expansion. I have also stressed that the expansion of Europe had been an important stimulus in its self-transformation. The European expansion then had various effects. Besides being a catalyst in the self-transformation of Europe, it has been the major process by which a worldwide international system has come into existence, that is to say, societal relations become globalized.

The European expansion does not only refer to subjugation of other societies or systems by the Europeans. It involves the tremendous impact upon the internal structures or characteristics of those societies. The expansion of Europe started with the Age of Discoveries from the mid-fifteenth century onwards. The Europeans discovered new lands and peoples they hitherto had had no knowledge of. However, before an examination of how Europe expanded and of its effects, we first need to answer the question of why Europe expanded. Why did the Europeans go into the oceans and not just remain within the boundaries of the Old World?

Why?

It is hardly possible for us to tell the exact causes that made the European explorers of the fifteenth and sixteenth centuries and their sponsors embark on oceanic voyages. We do not know what the particular motives were of Columbus in 1492 or Vasco da Gama in 1498. Yet, we could say that the state of the societies and economies, the rivalries and conflicts between the rulers, a policy dedicated to political and economical aggrandizement, the search for honour and glory, and an instinct for adventure could all have had their role.

Viewed within a perspective of world history, we can, however, propose another, and perhaps more pertinent, reason. The European expansion could be taken as the result of a previous expansion, namely the Islamic expansion. Ever since the emergence of Islam as a world religion and a system of society, Islam and Christianity had been in a fierce competition with each other. Furthermore, the rivalry between Islam and Christendom went on at the expense of the latter. Islam emerged in Arabia and quickly expanded all over the Middle East, most of Asia and North Africa. By the fifteenth century, Europe or Christendom was encircled by Muslims from Spain to the Eurasian steppes. The Islamic expansion culminated in the Ottoman conquest of Constantinople in the mid-fifteenth century. We could say that the Europeans were blocked on the East by the Muslims represented by the formidable power of the Ottomans. The Mameluks of Egypt had already established control over oriental trade with the fall of the last European colonies in Asia Minor at the end of the thirteenth century. The Ottoman stranglehold on the Levant ruined the Venetian trade. Moreover, the Ottoman navy was dominant in the Mediterranean until the second half of the sixteenth century. The principal gateway to the east and its riches was closed. Therefore, Europeans had to try the ocean.[38] That was what they did. The result was the rise of Europe to a central place in the world and consequently the establishment of a worldwide international system.

The European expansion, which was first directed towards the Americas with the discovery of the New World, included the Old World as well. It was so far-reaching that, by the end of the nineteenth century, the old, established civilizations of Asia and Africa were subjugated and incorporated within the European system. As Watson pointed out, the subjugation took place in the form of either direct or indirect imposition of European rule over the territories, for instance in India, Indonesia, Central Asia and Africa, or in terms of 'unequal treaties' as happened with China and Japan.[39]

European subjugation of the world was realized in the second half of the nineteenth century. It can, however, be said that the operations of the European international system were globally directed from its early beginnings in the sixteenth and seventeenth centuries. The fact that European rivalries came to be reflected in the colonial areas, and policies pursued in colonies and wealth transferred from colonies to Europe played a role in struggles between European states could be taken as an indication of the globalization of the European international system. It is widely agreed that, by 1850, European civilization clearly outstripped other civilizations and rose to world dominance.[40] Perhaps this can be taken as the distinguishing landmark between the strictly European international system and the worldwide international system in the history of the modern international system.

How?

How could such an expansion on the world scale become possible? In the expansion of Europe, the technical and military superiority of the Europeans is said to be the decisive factor. It was the Europeans who invented and made use of the sailing ship, not the Ottomans or the Indians. It was again the Europeans who developed an extensive and disciplined standing army of professional soldiers. Toynbee considers the modern Western sailing ship as 'the instrument and the symbol of the West's ascendancy in the world ... the vehicle that had created the possibility of world-unity in the literal sense of uniting the whole Human Race, throughout the habitable area of the planet's surface, into a single society'.[41]

In the process of European expansion, for Wright, gunpowder was essential. The expansion was made possible by imperialistic war. It is true, he argues, that missionaries and traders had their share, 'but always with the support, immediate or in the background, of armies and navies'. The European trade with American Indians in the seventeenth century, with East Indians in the eighteenth century and with China and Japan in the nineteenth century were initiated by armed forces.[42] Howard outlines the role of the military factor in European expansion as follows. The first era of expansion from 1500 to 1700 had been made possible by the sailing ship and with its guns. In the second era from 1700 until *circa* 1850, it 'was based on the organization and firepower of the disciplined professional troops developed by European states in the internecine conflicts they fought between 1660 and 1720'. According to Howard, while the steam engine and high explosives were determinant in the third phase after 1850, it was air bombardments that counted in the final stage from the 1920s.[43]

Gong argues that the expansion of the European international system should not be attributed simply to European superiority. At least in the beginning, it was not just a clash of military forces, but also a confrontation of the social and cultural systems. Yet, 'European military superiority,' he acknowledges, 'left non-European societies no choice, but to come to grips with the European standard of civilization'.[44] There is thus considerable agreement that European expansion was made possible by the technical and military superiority of the Europeans.

It would, I think, be a simplified view if we consider 'European superiority' only in military terms. Of course, it would be equally simplistic if it is understood in only moral terms. The Europeans were superior over others not just in a technical and military sense, but also as a result of the process of Europe's self-transformation examined above, in socio-political organization and economic capacity. The role of industrial production in European expansion may best be exemplified by the fact that 'by 1789, ... English mills using cotton grown in India and imported to England around the Cape of Good Hope were able to undersell the Indian handweavers in India itself'.[45]

Here, it should perhaps be added that one of the facilitating factors of European expansion, especially in the Americas, was the diseases which were transferred from the Old World to the New World and to which native populations had no inherited or acquired immunities.[46] This is by no means to underestimate 'the role of the military factor' as European states had the capacity to send armed forces to distant parts of the world in order to support their merchants. In the nineteenth century, the techniques and art of war available to the European forces were overwhelmingly superior to the ones held by others. That is why the British easily defeated the Chinese defenders of Canton.

Nevertheless, there was the other side of expansion. It has been said that in the same years there was an imperial decay in the Asian empires. Europe succeeded easily for the others were weak. For example, the Europeans came to India only after the disintegration of the Mauryan Empire. The Ottomans were defeated by the Europeans partly because it had been weakened by the Mongols and Safavis in the East from the fifteenth century onwards. McNeill concludes: 'Thus the extraordinary European world hegemony of the years 1850 to 1914 was ... a result of an accidental coincidence of Europe's new wealth and power with a period of exceptional weakness among Asian governments and ruling elite.'[47]

European self-transformation and expansion

I have noted that the expansion process and the self-transformation of Europe were interdependent. It could fairly be claimed that there was a mutual enhancement between the two phenomena. The expansion of Europe helped its transformation from the medieval system to the modern system. The self-transformation in turn facilitated the expansion process. For example, the consequences of the Discoveries made an enormous impact upon Europe and, in fact, upon the other civilizations. The price revolution, resulting from the flow of massive quantities of gold and silver from the Americas, caused high inflation. Price rises were dramatic in most of Europe, the Ottoman Empire and China. The spread of American food crops such as corn, potato and maize increased local food supply and led to population growth in Europe, Africa and Southern China.[48] That the capital accumulation, inflation and population growth had been the factors behind the crisis and turmoil of the sixteenth century out of which came the market economy, urbanization and industrialization is understood. The capital accumulation, which resulted from European expansion, made the industrial revolution possible which in turn made the expansion of Europe worldwide.

It could be said that the contact the Europeans established with the New World did not only bring capital accumulation which made the Industrial Revolution possible, but also contributed to the disintegration of the traditional institutions handed down from the Middle Ages. For instance, Chadwick argues that secularization in Europe was related to, among other events, 'the discovery of the true nature of other great religions and cultures of the world'.[49] In turn, one can rightly claim that secularization enhanced European expansion for it would be easier to establish relations with the other societies (especially those which had their own great religions) on a secular and rational basis rather than in religious (Christian) terms. Gong makes the point that as a parallel to the expansion of European international society into a global society, its identity changed from 'Christian' via 'European' to 'civilized' society.[50]

The most significant consequence of the expansion of Europe has been, of course, the formation of the modern international system, as we know it today. It has earlier been said that the principal characteristics of the modern international system originated in Europe. Hence, it means that those characteristics have either been exported by the Europeans to the world at large or adopted by non-Europeans. Historically, both of them could be observed, and they happened largely as a result of European expansion.

The traditional institutions which developed as a result of the relations among European states such as diplomacy, international law and the balance of power were transferred into the other zones of the world by the close of the nineteenth century. At first, the non-Europeans were not so sympathetic to, for example, resident diplomatic legations. Gong makes the point that China's sense of cultural superiority and self-sufficiency gave it little reason to send diplomatic delegations abroad, either temporarily or permanently. The Ottomans had more or less the same attitude.[51] Yet, non-European states admitted resident embassies of European states and later sent European capitals their own ambassadors, as they came to terms with the ascendancy of Europe.

A similar attitude can be observed with regard to international law. Accordingly, Barraclough noted that as the 'great movement of European expansion and encroachment in Asia and Africa reached its peak, the result, it was generally believed, would simply be [to] transpose the European balance of power, as it had developed during the past four centuries, from a European to a global plane'.[52] I have earlier said that many would-be challengers of the nation-state have eventually come to accept it. Similarly, at present, non-European states have come to accept the practices and institutions of the modern international system, although it is always possible to hear contrary voices such as the embarrassment felt by some with respect to Article 38 (1) of the statute of the International Court of Justice which directs the Court to apply the 'general principles of law recognized by civilized nations' in such disputes as are submitted to it.

In fact, the expansion of Europe did not only result in the transfer of some institutions to the states of America, Africa, Asia and Australasia. In most places, it actually created nations and states through direct or indirect rule. The traditional socio-political units of non-European zones of the Earth were either united or divided to fit in a nation-state. Those states which had long-established orders of their own were transformed into an order of nation-state. It was said that India had not been a nation before it was unified by the British. Bozeman argues that it was the English or European elite that reconstructed India's history, art and architecture; discovered its languages, religions and sacred texts; and identified the region's legal, social and political traditions in their full complexity. The West, she goes on, called forth Indian nationalism by giving this fragmented land a new sense of its own old cultural values and achievements.[53] The nation-state and international institutions have, as stated, become global. Yet, the transfer from Europe to non-Europe was not limited to nation-state and its institutions.

Europeanization

Equally important and impressive has been the exportation to, or adoption by, non-European societies of European or Western values. Values such as secularism, rationality, democracy, technology and science and, of course, nationalism became common among non-European peoples as well; not that they have been adopted by all societies or all segments of societies, but they have attained a high saliency and their impact has been pervasive. The creation of a Europeanized or Westernized elite throughout Asia, Africa and Oceania in the nineteenth century was perhaps the main factor by which Western values found ground in the non-European societies.

The expansion of Europe pervaded the traditional values of non-European societies. The process of Westernization or Europeanization, whether it is promoted by the Europeans or self-consciously adopted by the non-Europeans, caused an identity crisis within non-European societies, all of them having been divided between those sections of society, especially the elite, heartily adopting and arguing for a secular, Westernized order, and those who see the preservation of the traditional values of society as the best course. The following remarks by Nehru, the man who was one of the leading figures of the Third World or the non-aligned countries, show how effective European values were among the elite: 'All predilections (apart from the political plane) are in favor of England and the English people, and if I have become what is called an uncompromising opponent of British rule in India, it is almost in spite of myself.'[54]

According to Gong, the response of non-European societies to the European expansion has been, initially, to preserve 'the traditional character of society while providing the technological means necessary for its defense', and later, to isolate 'the requirements of the standard of "civilization" which contributed to universal values and industrial growth from those which would lead to wholesale Westernization'. Gong observes that the second effort continues today.[55]

This brings us to the issue of the response of non-European societies to the expansion of Europe, or of the relation between the Europeans and non-Europeans. Gong's observation provides us with some clues. We see that the globalization of the characteristics of the modern international system from its European basis has not been through simple imposition by the Europeans. This is in no way to deny the European missionary zeal in the course of the expansion. However, after the initial period of indifference and sense of superiority, non-European societies tried to make a distinction between the 'technical aspects' which were seen as the causes of European ascendancy, and social and cultural values which were proper European. Then, as Gong observed, an effort goes on to distinguish 'universal values' from those which have been understood as exclusively European. In other words, the non-Europeans did not object to the 'technical', 'universal' tools and values. They did not reject industrialization, though this was, too, largely of Western origin. Here, it is assumed that it is possible to make a distinction between those two categories and it is also assumed that the adoption of the first is not detrimental to the societal values of a traditional character.

Those assumptions and efforts have some truth in them. As I have stated in this book, social identifications are not static, closed, unchanging and ongoing as they are. Furthermore, it could be said that every collective entity has some exclusive and some non-exclusive identificational characteristics. That is how societies could use the values and techniques of the other societies and still preserve their individual identities. I have earlier made the point that interactions between civilizations, and hence societies, have not always been even. When one civilization becomes the attraction centre of the world, it is natural to expect the others to adopt some characteristics of the former. It is the dynamic, historical, changing character of the societal identifications and the fact that they are not absolutely exclusive which make such distinctions and, paradoxically, the transcendence of these distinctions, possible. This is what accounts for the emergence of the modern international system, which is global, inter-civilizational and trans-civilizational.

It is global, as it comprises the entire globe. It is inter-civilizational for there is a sufficient degree of interactions among civilizations. It is inter-civilizational since the values and characteristics that define civilizational identities have not yet faded away. Perhaps one reason for the multi-civilizational character of the modern international system can be found in its European origins. The modern European international system was, from the beginning, based upon the agreement and coexistence of differences. The existence of separate and different types of states was the basis of the system. Once the existence of different states is accepted, it is one step further to accept the existence of different civilizational entities.

The modern international system is, however, Western-centred, simply because it emerged as a result of the expansion of Europe and with the ascendancy of Western civilization. From the sixteenth and seventeenth centuries on, Western civilization rose to a central position among the existing civilizations. It is trans-civilizational, as international or inter-societal interactions do not take place along only civilizational lines, but are articulated through multiple channels. The state, though a prominent unit, does not have a monopoly on the channelling of inter-societal relations. With the degree of speed and improvement of the techniques of communication and transportation, there emerges a higher degree of plurality in terms of inter-societal relations than the one associated with national or state lines.

Having examined the emergence and development of the modern international system as such, there are still some further points to be dealt with. The next, and final, chapter takes issue with them.

7
Modern International System II or the White Man's Burden

For a more adequate understanding of the modern international system, one needs to clarify when the system began to emerge in Europe and when it became worldwide. From a perspective of universal history or world history, the search for the exact origins of the system in Europe may be considered as a futile endeavour, as there could be, and indeed there were, worldwide events and processes which determined the emergence of the modern international system. Yet the search for the origins of the system enables us to see it not only in historical terms, but also in comparative terms.

It has already been noted that the emergence of the modern international system in Europe occurred in interaction with the non-European world. When the issue is considered in relation to the non-European world, one may argue for an international system, which was broader than the European international system and other uni-civilizational systems, and which preceded the modern international system, namely (following Hodgson and McNeill) what I call the 'Afro-Eurasian international system'.

In this regard, we must examine the role of civilization in and its significance for the modern international system. The modern system has the nation-state as its principal unit and the nationality or national identity as the primary focus of collective social identification. I have already shown the prevalence of the national identity in the modern world. Nevertheless, one cannot argue that civilization as another form of collective social identification has disappeared. Civilization as a form of social identification operated in the modern international system throughout its history. To deal with these issues, I start with the question of when the modern international system began.

When?

So far, I have freely averred that the modern international system came into being in the sixteenth century. I have, yet, avoided giving a precise date. If the history of the system could be divided into two periods, one being the period of the European international system and the other referring to when that system became worldwide, then, it seems that one needs to identify two starting points. However, discussions in the literature are concentrated on when the modern European system began. This question has been closely associated with the question of when the 'modern history' commenced. The beginning of the system is associated with the beginning of the modern period, because it is through those events and developments seen as forming modernity that Europe, it is argued, assumed its principal characteristics, some of which also laid the basis for the modern international system as shown in the previous chapter.

Not surprisingly, there is a controversy around the beginning date of the system. Most scholars think that the modern history or the modern international system in its European era originated at the end of the fifteenth century and in the beginning of the sixteenth century. While some identify exact starting points, some avoid giving a specific date.

The conventional verdict

For Bolingbroke, the modern period started at the close of the fifteenth century. The Göttingen School agreed with Bolingbroke, in taking the great discoveries as the dividing line between medieval and modern times. In their view, for instance, for Heeren, the discovery of America, the invention of gunpowder and printing and the discovery of the new route to the East Indies make the close of the fifteenth century a convenient dividing line. French historian Michelet was more specific: the year 1494, when Charles VIII of France invaded Italy, could best be taken as the beginning of the modern period.[1] Ranke joins him in considering the invasion of Italy by the French and Spanish kingdoms to be the beginning of the modern European system.[2] Lord Acton also shares the view that the close of the fifteenth century serves as the marking line between the modern and the medieval period. In his view, the discovery of the New World and the recovery of the Old World with the Renaissance were the distinguishing landmarks. By AD 1500, peculiar characteristics of the modern state had emerged and European nations had a full measure of differentiation. He argues that the Italian wars determined the main conditions of modern politics such as balance of power.[3] Dehio forcibly makes the point that the modern European states system 'came into existence at a quite definite moment: the beginning of the struggle among the great powers over Italy in 1494'. Modelski, following Dehio, starts his 'Long Cycles of World Politics' in 1494.[4]

Not all scholars are, of course, keen on precise dating. Guizot saw the signs of preparation in the fifteenth century and he held that 'it was with the sixteenth century ... that modern society really commenced'.[5] In the introduction to *The Cambridge Modern History*, the editor took the Renaissance to mark the beginning of modern history. The emergence of the competing nations instead of a European commonwealth during the fifteenth and sixteenth centuries was considered the starting point for modern history.[6] Bowle, too, agrees with the view that emphasizes the role of the Renaissance, the Discoveries and the Reformation and believes that 'the making of the framework of the modern world' happened in the sixteenth century and extended into the seventeenth century.[7] McNeill, stressing the significance of the Discoveries and the following developments, especially the Reformation, takes the year AD 1500 as the dividing point between modern and pre-modern times.[8] Finally, Bull and Northedge give importance to the fifteenth century and the sixteenth century as the beginning of the system. While the former thinks that 'the European dominated international system' emerged in the sixteenth century, the latter places the origins of the system some four hundred years ago, if *raison détat* is thought to be the most distinctive mark of the modern government.[9]

As seen, though they may pay attention to specific dates or particular events, the scholars seem to agree upon the end of the fifteenth century and the beginning of the sixteenth century as the period when the modern European international system emerged. Behind this dating has been the view that the Renaissance, the Discoveries and the Reformation were the developments which set in motion the principal characteristics of the modern European international system. Strangely enough, there is no precise dating of the Renaissance in spite of the significance attached to it. In his seminal work, Burckhardt traces the signs of the Renaissance back to the poems of the unknown 'Clericus' of the twelfth century.[10] Though the twelfth century is a long way back from the conventional date, it seems that it has been a conventional view to start the modern international system at the turn of the fifteenth century. There are, however, those who dissent from this conventional view and take the origins of the system further back or ahead.

The dissenting voice

From the conventional date of 1494, Wight seeks 'the rudiments and premonitions of the states-system' further back. He finds it in the Congress of Mantua (1457–60), which was the first pan-European gathering, and in the Peace of Lodi and the Most Holy League of Venice (1454), which founded the Italian Concert and the first system of collective security. Wight finds the dividing line between the medieval international papal monarchy and the modern secular sovereign states-system in the Council of Constance (1414–18), 'which is as far back as one need go in the search for the origins of the states-system'. Though he sees the breakdown of the *Respublica Christiana* in the beginning of the fifteenth century, he states that the newly emerging system matured only in the mid-seventeenth century with the Peace of Westphalia. He makes it clear: 'At Westphalia the states-system does not come into existence: it comes of age.' Wallerstein, somewhat evoking Wight in his analysis of the creation of the 'European world-economy', considers the years 1450–1640, what he calls 'the long sixteenth century', to be the 'meaningful time unit'.[11]

What we see in Wight's argument is that, in terms of the origins of the modern international system, the important point is the breakdown of medieval universalism which happened with the emergence of separate units. Those who argue for the conventional view, on the other hand, seem to give significance to the character of the new system besides the breakdown of medieval universalism. For Wight, I suspect, it does not have importance in the first degree if the emerging units have a particular character, say, secular or non-secular.

Unlike Wight, others who dissent from the conventional view argue that the system emerged much later, sometime at the end of the seventeenth century and in the beginning of the eighteenth century. The most prominent advocate of this view is Hinsley, according to whom the modern 'European states' system emerged in the eighteenth century, and not at an earlier date'. Hinsley argues that the present-day structure of international relations is a structure of great powers, not just one great power, and happened in the early eighteenth century when the France of Louis XIV was stopped by a coalition of powers. The new era, for him, represented a change of kind, not merely of degree.[12] Hinsley seems to have taken the treaties of Utrecht (1713), the last treaties referring to *Respublica Christiana* and also the first treaties declaring that they were made in order to preserve the European balance of power, as showing the early eighteenth century to be the beginning of the system.

Hinsley's account, which was described as a variant of the Westphalian interpretation by Wight, appears to rely on the idea that the emergence of separate units is not sufficient for the system to come into being, but it is required that none of the units should attempt, or be in a position, to establish a hegemony within the system. Needless to say, post-eighteenth century history of the modern international system involves sufficient attempts for hegemony, from Napoleonic France to Hitlerian Germany and all the talk about American hegemony in the aftermath of the Second World War.

Among the others arguing for the late seventeenth century and the early eighteenth century are Butterfield and Barraclough. Butterfield regards the scientific revolutions of the seventeenth century as forming the foundations of the modern period.[13] Barraclough expresses a similar view in declaring that 'the great break in outlook, the change in intellectual climate ... came, not with Renaissance or Reformation, but with Enlightenment which set the course for the developments of the nineteenth and twentieth centuries'. More specifically, he sets the start of 'modern' history around the years 1660–80.[14] Northedge, too, suggests a similar view. If the system is thought to be dominated by the national interest, he inclines to regard the beginning of the nineteenth century as the inauguration of the age of the modern international system.[15]

Behind this third interpretation, we see the view that puts emphasis on the Westphalian settlement, the scientific revolution, the emergence of the great powers, the development of international law, the professionalization of diplomacy and so on. The view that the modern international system should be taken to have started with the Westphalian settlement of the mid-seventeenth century has been a credo for most students of mainstream International Relations.[16]

Origins of the global system

So much for the origins of the modern European international system. How about the origins of the global international system? Stated in another way, when did the modern international system, which originated in Europe, become worldwide? I have already stated McNeill's observation that by the second half of the nineteenth century European civilization outstripped all the other civilizations. It has also been noted that the emergence of the modern European international system and its expansion went hand in hand. It is widely agreed that the expansion of Europe embraced the whole world by the end of the nineteenth century. In other words, by the twentieth century, the whole world was brought into a single system. The striking landmarks are given as the emergence of non-European states and their participation in the European network of conferences or diplomatic relations such as the Paris Peace Conference of 1856, and the Hague Conferences of 1899 and 1907. In the twentieth century, it is said that, with the end of European dominance, a truly global and worldwide international system has come to be established. The significant landmarks are here given as the rise of non-European great powers, the two world wars and de-colonization.[17]

Having thus surveyed the literature on the question of when the modern international system came into being, I would like to note some points. First, we see that when the system is considered to have emerged depends heavily upon how it is perceived and identified, and upon what characteristics of the system are considered as primary and central. Second, in the literature we cannot find a commonly agreed and definite starting date. This follows from, on the one hand, the first point and, on the other hand, the futility of the search for origins. It is not easy to find a definite date for the beginning of the system because there can hardly be one. Even if one is suggested, as some did, it is bound to be arbitrary. The reason can be found in the arguments advanced in the earlier chapters of this book, namely the arguments on the unity and continuity of history and social process. One can always find the influence of what has been on what is happening or will happen. To recall Graham Greene, 'a story has no definite beginning and end'. Whenever we want to trace the earliest sign of a phenomenon, history takes us further and further back.

It is possible to find the traces of the defining characteristics of the modern (European) international system in the periods earlier than the sixteenth century or the end of the fifteenth century. In fact, it has already been pointed out by Wight that one could go from the end of the fifteenth century to the beginning of the fifteenth century. He even takes it further back. Though the real break between medieval universalism and modern particularism became apparent in the fifteenth century, it was developing through the fourteenth century.[18] It is already recorded that rival powers to the would-be-universal authority gained recognition in the beginning of the thirteenth century.

McNeill has already told us that the origins of the bureaucratic form of government and standing professional armies could be followed further back to the ancient Near Eastern systems. It is a commonplace view to see a strong similarity or affiliation between the modern (European) international system and the classical Greek city-states system. Bozeman thinks that many of the institutions associated with

Athenian democracy may have been adaptations of the earlier institutions that were in operation in the Middle Eastern civilizations. The secret ballot, for instance, was probably used by the Sumerians and Indians long ago. Above all, the very institution of the city-state had been known in the Near East.[19] Hume showed us that the idea of the balance of power had existed in classical antiquity. Thucydides expressed this in explaining the outbreak of the war: 'the real cause I consider to be the one which was formally most kept out of sight. The growth of the power of Athens, and the alarm which this inspired in Lacedaemon, made war inevitable.'[20] All this shows the futility of seeking a definite beginning point. Perhaps, that is why most scholars avoid giving a certain date and prefer to speak of a time-span such as 'the fifteenth century', 'the beginning of the sixteenth century', 'by the seventeenth century' and so forth.

Afro-Eurasian international system

The difference between the modern (European) international system and the medieval European system has been well scrutinized. However, it is argued that there is an affiliation between the two systems. According to Bozeman, the millennium between AD 500 and AD 1500 constitutes the first great chapter of a narrative analysis of the modern international system. It would be said that the present international relations have their anchorage in the European region and in the millennium from AD 500 to AD 1500. For example, the Crusades, by establishing contacts and exchanges between hitherto rather isolated civilizations, made it possible to formulate the concept of a world community.[21] Bozeman's observation is crucial. It does not just show the existence of the rudiments of the modern (European) international system, but it brings us to another subject. By the formulation of a concept of world community via the contacts and exchanges between hitherto isolated civilizations, it shows us the existence of an international system preceding the modern European system; an international system which is wider than the medieval European system, and which is as inter-civilizational as the modern international system. Earlier, in Chapter 5, I quoted from Wight, three examples of inter-civilizationally international systems other than the modern international system. Was there another one?

It could be argued that there was an international system, albeit with a lower degree of systemness in comparison to the modern international system, extending beyond Europe and possibly covering the whole of Eurasia. It might be called the 'Afro-Eurasian international system'.

Bull and Watson in the introduction to their edited work, identified four regional international systems other than the one in Europe when the latter began to expand: the Arab-Islamic system stretching from Spain to Persia, the international system of the Indian subcontinent and its extensions eastward, the Mongol-Tartar dominion of the Eurasian steppes, and the Chinese system.[22] The first thing that should be noted about this list is that all the regional, not global, international systems take place in Eurasia and at least three civilizations correspond to them. At this point, we may naturally ask: Was it the case that these four systems sharing the then known world remained as closed systems without any interaction among them, or that they had interactions with each other? Based upon the historical record, we are, I think, justified to assert that these so-called regional systems were in interaction so as to be part of a single system, that is, the Afro-Eurasian system.

An inter-civilizational system

The Afro-Eurasian international system was an inter-civilizational system in the sense that the civilizational identity as a form of collective social identification was salient and the inter-societal exchanges were largely channelled through civilizational lines. It was also a multi-civilizational system in the sense that it comprised multiple civilizations. The system included the major Eurasian civilizations, namely Europe (Christians), the Middle East (Muslims), India and China.

We have been told by Hodgson and McNeill that there were interactions among different societal, or regional, or civilizational entities long before the modern period. We know that there were exchanges between distinct civilizations by 500 BC when the Middle Eastern civilization, according to McNeill, had preponderance in the then world. From 500 BC to AD 1500, we do not see any single civilized centre enjoying a definite preponderance. After AD 1500, the European centre assumed predominance. However, the fact that there were inter-civilizational exchanges does not mean that we can necessarily place them in one system.

McNeill's description of the interactions and interdependencies among the civilizations of the Old World allows us to speak of a then worldwide system that he calls the 'Eurasian-African ecumenical system'. His remarks explaining its existence are worth quoting at length:

> The reason was that mercantile practice had, in fact, slowly created a workable code of conduct that went a long way towards standardizing

> encounters across cultural boundaries. Even the arcanum of religion made room for outsiders and unbelievers, since the principal religions of the Eurasian world – Christianity, Confucianism, Buddhism and Islam – all agreed in exhorting the devout to treat strangers as they would wish to be treated themselves. Thus, despite the fact that no single set of rulers had ever exercised political sovereignty across the whole Eurasian-African ecumene, a bare-bones moral code did arise that went a long way towards reducing the risks of cross-civilizational contact to bearable proportions. Little by little across the centuries, local rulers of every stripe learned that they could benefit mightily by taxing instead of plundering strangers ... As these attitudes became general, so that an enforceable (and remarkably uniform) merchant law arose in the ports and other great urban centers of Eurasia, and was supplemented by an informal body of customs for dealing [with] strangers that extend into the rural hinterland, the structure of the ecumenical world system approximated very closely to that of the separate civilizations embraced within it.[23]

What is so striking in this quotation is that it informs us about the nature and extent of the relations among societies of the Old World which allow us to confidently argue for, what I have called, the Afro-Eurasian international system. We see that the interactions among those societies involve a workable code of conduct, standardization of encounters, a moral code, an enforceable and uniform merchant law, and an informal body of customs, all of which are the elements we associate with international systems. It is thus such a system of interactions that means those societies may be said to amount to an international system. There emerged a degree of interdependence that international systems are generally observed to have.

I have earlier noted that, by AD 1000 local civilized societies of the Old World began to have interactions and exchanges, and from AD 1500 the peoples of the Americas and Australasia were incorporated into this network of interactions. We have been told by Southern that there was already an international trade between Latin Christendom, Constantinople and the Islamic World by the eleventh century. He even goes further and makes the point that the restoration of the Mediterranean in European politics in the late twelfth century is one of the main determinants in later Medieval Europe.[24] The international trade between those three civilizations, in fact, extended into India and China as well. All this suggests that it is possible to speak of an Afro-Eurasian international system, comprising Europe, the Islamic world,

India and China, before the modern international system which embraced the Americas and Oceania as well.

The Afro-Eurasian international system, as it is named here, was a system in the sense that the behaviour of one part of it has been a factor in the calculations of the others. Or, some degree of interdependence among them existed. This could best be seen in the relations of the Islamic world and the Christian world. From the eight or ninth century onwards, the two realms were a constant determinant taken into account by each other. The Crusades were one of the best indicators of this relationship.

That there was a degree of interdependence between the Eurasian societies could also be seen in the worldwide impact of the Discoveries. As noted, the Discoveries affected, besides Europe, the Ottoman Empire and China. With the increase in the supply of gold and silver in Europe via transfer from the Americas, price rises occurred not only in Europe but also in the Balkans, the Middle East, India and China. Similarly, the introduction of the new crops such as potato and corn led to population increases in Europe and non-European societies from the Middle East to China. We could not expect a devastating impact on the other societies of the European discovery of the New World unless, of course, there had already been some relations between them involving mutual dependency.

Similarly, the existence of an Afro-Eurasian international system before the modern international system would, I think, be one of the factors that count for the negligible Europeanization in Asia and Africa, compared to the one achieved in the New World, in the age of European expansion. The degree of Europeanization in the societies of the Old World has been rather slim, as they had long been in interaction with the Europeans and had already been taking part within the same system with the Europeans. In many respects, those so-called regional systems of the Old World were dependent upon each other. More specifically, we observe this interdependence in the relations between the Ottoman Empire and the emerging European states system.

Ottoman Empire and the European states-system

From its emergence as a power in the fourteenth century, the Ottoman Empire expanded at the expense of Europe. We can conveniently consider it as an imperial system from the first half of the fifteenth century. It occupied, controlled and administered one-quarter to one-third of the European continent from the fourteenth century to the nineteenth century. As already shown, the modern European states system is conventionally said to have emerged from the late fifteenth century onwards and consequently the Ottoman Empire was in Europe when the European system began to come into being. From its emergence as a formidable power, the Ottoman Empire had been a continuous consideration for the Europeans. So, the modern European states-system was never isolated from the other systems. The Ottomans played a major part in the formation and working of the European international system and this shows that a process of mutual dependence operated between the two systems.

As early as the first stage of the Italian Wars from 1494 onwards, the Ottoman Empire was an important actor in the Italian system that traditionally was seen as the forerunner of the modern European system. The Italian courts maintained diplomatic relations with the Ottoman Sultan. When the so-called Second Holy League was signed in 1495 with almost Europe-wide participation, not just by the Italian states, Mattingly tells us, the Turkish Ambassador was present, as an observer in the signing ceremony. The New League is said to have transformed the Italian system into a European one.[25] The Ottomans took place as an active party in the second stage of the Italian Wars. Just as Ottoman engagement in the rivalries between the Italian states and the intervening states in the Italian Wars led to a pan-European gathering with the Second Holy League, it can rightly be said that the struggle between the Ottomans and the Habsburgs throughout the sixteenth century linked the two European systems, the Southern system centred in Italy and the Northern system comprising Sweden, Poland-Lithuania and Muscovy.[26]

The historical record shows that the Ottoman Empire became an active participant in the emerging European balance system. The Sultans pursued a conscious policy of balance *vis-à-vis* the European powers so that the rise of the nation states was to a certain degree facilitated. Similarly, the European sovereigns took into account the Sublime Porte in their calculations of the balance in Europe and did not hesitate, from time to time, to align with the Sultan against each other. According to Dehio, the Ottoman Empire became a counterweight to the unifying tendency represented by Charles V. The introduction of the Empire into the European balance-of-power system and European diplomacy played a most significant part in preserving the freedom of the system of states.[27] In 1532, Francis I admitted to the Venetian Ambassador that he saw in the Ottoman Empire the only force guaranteeing the continued existence of the states of Europe against Charles V. Indeed, in 1536 we see that this guarantee was in some sense given with the attempt at a Franco-Turkish Treaty, which is said to have provided the Europeans with a model in their relations with the Asian empires later in terms of unequal treaties.

The role of the Ottoman Empire in preserving the European balance and thus nation-states can be seen to continue later in the support and encouragement given to the English and Dutch in the period after 1580 when these nations proved to be the champions of European resistance to the Habsburgs' attempts at hegemony. In the sixteenth and seventeenth centuries, support for the Protestants and Calvinists was one of the fundamental principles of Ottoman policy in Europe. The Ottoman pressure on the Habsburgs was an important factor in the spread of Protestantism in Europe. The Westphalian settlement, allowing for the coexistence of multiple sovereign states, became possible through this pressure on the Habsburgs as observed by Watson: 'The Habsburg bid to establish a hegemonial system in Christian Europe was defeated, decisive Westphalian formulation of the anti-hegemonial nature of the European international society was made possible by the Ottoman pressure on the Habsburgs.'[28]

In terms of trade relations, we can observe that both the Europeans and the Ottomans took the other into account. The Ottoman Empire pursued the balance policy in its trade relations with the Europeans, notably in terms of the Capitulations. In order to prevent the dominance of one state in the Levant trade they always favoured the rival nations. Against Venetian dominance, they supported first the Genoese, then the Ragusans and then the Florentines in the fifteenth century. In the sixteenth century the French took the lead, and in the seventeenth century came the English and the Dutch.[29] In short, the Ottoman Empire was a significant force in the European balance-of-power system from the fifteenth century to nearly the end of the seventeenth century, the formative centuries of the system.

Both contemporaries and scholars indeed recognized that the Ottoman Empire was in, and essential to, the European balance system and there was a mutual dependence between them. As already noted, in the early sixteenth century Francis I admitted that the Ottoman Empire was the only force to prevent the emerging states of Europe from being transformed into a Europe-wide empire by Charles V. In the late sixteenth century, Queen Elizabeth I opened relations with the Ottoman Empire. One of the motives of the Queen was certainly the expansion of trade and the second motive was the idea that the Sultan could balance the Habsburgs in the East and consequently relieve Spanish pressure on her. Elizabeth I even stressed that Protestantism and Islam were equally hostile to 'idolatry' (Catholicism). In granting Capitulations to the English and the Dutch, the Sultan, too, considered that these nations were the champions of the struggle against the idolaters.[30] In the late eighteenth century, the place of the Ottomans in the European balance system was acknowledged in the British Parliament. Similarly it has been reported that Catherine the Great of Russia explicitly recognized it.[31] The scholars too considered the Ottoman Empire within the European balance system from the Renaissance onwards.[32]

What one can draw from the foregoing analysis of the mutual positions of the emerging European international system and the Ottoman imperial system is that the systems were closely interwoven and they were in constant interaction. I have already shown their interdependence in terms of European expansion. This analysis of the mutual dependence between the Ottoman Empire and the European international system in its formative (and, of course, later) centuries unequivocally leads us to conclude that, *pace* Bull and Watson, the so-called regional systems of the Middle East and Europe were not isolated from each other. They had frequent relations with each other and the nature of these relations was not always warlike. The Europeans and the Ottomans did not always aim at plundering each other, they were not in a permanent state of war as the orthodox understanding of the *Christendom* versus *non-Christendom* or *Dar-al-Harb* versus *Dar-al-Islam* dichotomy would have us believe. They do not seem to constitute two antagonistic systems, but to be parts of a greater system, that is, the Afro-Eurasian system, together with the other societies and civilizations of, what Hodgson calls, the Afro-Eurasian zone.

Inter-civilizational and trans-civilizational system

Viewed this way, the formation and development of the modern international system wears a different aspect. It is not just an outcome of the expansion of Europe, but rather the result of the transformation of the then existing Afro-Eurasian system together with the inclusion of the Americas and Australasia. Buzan and Little are right in arguing that in order to understand the contemporary global international system one must examine the wider history of the Afro-Eurasian system rather than tracing the origins of Europe.[33] One may even argue that Europe's self-transformation was not a European event; it was not just Europe's own doing. As Hodgson put it, 'without the cumulative history of the whole Afro-Eurasian Oikoumene, of which the Occident had been an integral part, the Western Transmutation would be almost unthinkable'.[34]

Nonetheless, we cannot deny that the European expansion and Europe's self-transformation have been the major piston in the transformation of the Afro-Eurasian system and in the incorporation of the New World, leading to the formation of the modern global international system. In the transformation of the Afro-Eurasian system, it was through the self-transformation of Europe that the principal focus of the social identification of the modern international system, that is, nation-state, emerged, whereas the Afro-Eurasian international system had its principal social identification in the civilizational unit. Moreover, it has been with the expansion of Europe that the characteristics of the modern system have become world-wide. In the incorporation of the New World, this was simply because the incorporation of the societies in the Americas and Australasia into the Afro-Eurasian international system was realized through the expansion of Europe.

Unlike the Afro-Eurasian system, in the modern system, as stated, the national identity has become the major social identification. We see a decline in the effect of civilizational identity as a result of the loosening of civilizational ties. Yet, it does not mean that civilizational identity has been fully replaced by national identity. In fact, civilizational identity has been a persistent factor in the development of the modern (European) international system. If the development of the modern international system is a result of the emergence and struggle of nation-states on the one hand, it is accordingly a result of the struggle and interactions of civilizations, on the other. This leads us to the question of to what extent the civilizational identity or civilizational consciousness played a role in the development of the modern international system.

The role played by civilizational identity can best be discerned in the process of European expansion. Europe, albeit in different names and different extensions, has been a major centre of civilization. There is a European civilization.[35] The question is whether the Europeans had (or have) a civilizational consciousness in their relations with the non-Europeans. The existence and influence of a civilizational identity and consciousness can be observed both through the policies of each European state and through the collective actions of European states or those states that would be associated with the European civilization.

White man's burden

It has been said that nearly every civilization considered itself as *the* civilization and others as uncivilized societies. Europe has been no exception to this, perhaps inherent, tendency. As a reflection of this, one finds some kind of missionary zeal in the name of civilization in the European expansion, expressed in Kipling's famous poem, 'White Man's Burden'.[36] In fact, the expansion of Europe into the outer world was justified in terms of the idea that civilization should be spread all over the world. In the end, all peoples were to be civilized. In 1798, as Napoleon set off for Egypt, he shouted to his troops: 'Soldiers, you are undertaking a conquest [with] incalculable consequences for civilization.'[37] Similarly, Vincent tells us that 'prominent in the British justification of empire was the idea of trusteeship over backward races, who could and should be civilized and educated'.[38] In this missionary zeal for civilization in the modern European states, one could find an echo of Christendom and the spirit of the Crusades.[39] This missionary zeal, the idea that they are working for the spread of civilization, appears in the policies of nearly all European states, especially, when they are dealing with non-European societies. Of course, the civilization to be spread was the European or Western civilization. However, it was presented, not just as *a* civilization, but as *the* civilization.

The missionary zeal for civilization and civilizational consciousness has also been effective in the collective actions of the Europeans. In fact, it would not be wrong to say that it was through the existence of a common civilizational identity and its consciousness by the states that Europe was able to act collectively. From the start, as already stated, the states of Europe considered themselves as members of a civilization, distinct from the others. As Northedge and Grieve stated, the Europeans or European powers had a European self-consciousness, based on the understanding that 'they were European, Christian, civilized and white, and thus enjoyed a common stock of traits and values which separated them from the rest of the world'. In response to this, it is argued: a sense of 'Asianness' emerged. Panikkar spoke of a 'common feeling of Asianness' among the Afro-Asian Third World countries and attributed it to the earlier emphasis the European states placed on their Europeanness in dealing with the Asian countries.[40]

That a 'standard of civilization', through which non-European states were to be tested, was formed by European states is a plain expression of the civilizational self-consciousness. Gong's work is undoubtedly one of the most comprehensive and penetrating on the standard of civilization. He outlines the requirements of the standard of civilization as understood by the nineteenth century European elite and statesmen as follows. The standard required that a civilized state: (a) guarantees basic rights (life, dignity, property, and freedom of travel, commerce and religion); (b) exists as an organized political bureaucracy with some efficiency and capacity to run the state machinery and capacity for self-defence; (c) adheres to generally accepted international law and maintains a domestic system of courts, codes and published laws to guarantee legal justice for all within its jurisdiction; (d) fulfils the obligations of the international system by maintaining adequate and permanent avenues for diplomatic interchange and communication; and (e) conforms to the accepted norms and practices of the 'civilized' international society. However, the standard was more than a mere list of requirements. It had an implicit subjective part of 'unspoken assumptions' such as the 'instinctive reactions, traditions, and modes of behavior'.[41] According to Gong, 'the standard of civilization' defined the internal identity and external boundaries of nineteenth century international society.[42]

The collective identity of the Europeans, based upon a common civilizational consciousness, showed itself in the European attitude towards the non-Europeans. The European powers are said to have at all times adopted a different and much lower standard of conduct towards the extra-European world than that which they followed in their relations *inter se*. They continued to operate against the colonial peoples in the Americas and Asia in ways that were not permissible in Europe between states not formally at war. 'The unequal treaties', which were justified on the basis of the claim that they were for the provision of protection which every civilized state offered, were another indication of the effect of the civilizational identity and consciousness. The common attitude of Europeans against non-Europeans even went further and ended up in a collective action. The intervention against the Boxer rebellion was undertaken by a concert of 'civilized' powers, including the United States and Japan, in the name of civilized society. When the question of the continued existence of extraterritorial jurisdiction came up, the Europeans stood together against Japan.[43]

Civilization and the Terrible Turk

That the Europeans had a common civilizational consciousness and it affected their dealing with the non-Europeans may best be followed through the relations between the European states system and the Ottoman Empire, or 'civilized Europe' and the 'Terrible Turk'. As I have already emphasized, the Ottoman Empire was in Europe when the modern European states system began to take shape. From the very beginning the two systems had been in close contact and experienced interactions involving not just conflicts and war but agreements, alliances and commercial exchanges. For half a millennium, The Empire controlled and administered one-quarter to one-third of the European continent. 'The logical conclusion ought to be,' Naff points out, 'that the Ottoman Empire was, empirically, a European state. The paradox is that it was not. Even though a significant portion of the Empire was based *in* Europe, it cannot be said to have been *of* Europe.'[44] Despite the establishment of formal relations between the Great Queen, Elizabeth I, of England and the Sultan in the late sixteenth century, when the British Parliament raised the question of whether Turkey was within the European balance, it was deemed to be within the balance-system *in some respect*. It was only with the Paris Peace Treaty of 1856 that the Sublime Porte was formally admitted into the Concert of Europe. Nevertheless both the provisions of this Treaty and the legitimacy of the Ottoman entry into the Concert of Europe have been subject to frequent discussion.[45]

Despite the existence of extensive relations so as to form 'a pattern of alliance' between Europe and the Turk, the fact that the Sublime Porte was not formally accepted into the European system, and even this went on being questioned, clearly proves that not the explicit provisions but the unspoken assumptions, or not just the 'logic of *raison détat*' but the 'logic of culture' were effective[46] because the Turk did not share the common European identity. The general public perception had it that the 'Terrible Turk' was unspeakable. The Ottoman Empire was used as a unifying element and it was the significant 'other' of the Europeans.

The Turk has not been considered as being of Europe for he has been the major threat to Europe, or at least has been perceived as such. The Turk was then, to use the present-day expression, 'otherized', being described in negative terms and thus affirming European identity. Apart from the physical existence of the Empire in Europe, the religious divide has been the basic factor behind the negative perception of the Ottomans. For the Europeans Islam was just a heresy and Muslims were not trustworthy. While Martin Luther regarded Islam as a 'movement of violence in the service of the anti-Christ' which is 'closed to reason', Voltaire portrayed Mohammed as a 'theocratic tyrant'. Ernest Renan dismissed Islam as incompatible with science and a Muslim as 'incapable of learning anything or of opening himself to a new idea'.[47] Such characterizations were extended into the representations of the Turk.

The Turk was predominantly portrayed in pejorative terms: seen as the terror of the world by the Elizabethan historian Richard Knolles, and as incapable of feeling friendship to a Christian by Paul Rycaut.[48] According to Rousseau, Turks were the barbarians who conquered the civilized Arabs.[49] Burke told the House of Commons that the Turks were 'worse than savages' and that 'any Christian Power was to be preferred to these destructive savages'.[50] When classifying humanity in three categories of the civilized, the barbarian and the savage, James Lorimer defined two groups of the progressive and non-progressive under savagery. Of the Turks, he said that they did not even belong to the progressive races of humanity.[51] In the view of Sir Charles Elliot, the Turks have been a destructive force; they destroyed a great deal but constructed nothing.[52] There is no need to extend the examples.[53] With all these characterizations and representations, the Turks have been used to serve for the affirmation of the Europeans. The Turk was then the 'perfect barbarian' for the Europeans in order for them to readily affirm civilized Europe.

Nevertheless, as I have already said, there were permanent and extensive interactions between the European states and the Ottoman Empire. The Sublime Porte exchanged envoys and ambassadors with the European courts as it was the only non-European court accepting or having European resident ambassadors by the nineteenth century and it sent temporary envoys to the European courts frequently for long periods until the end of the eighteenth century when it began to reciprocate resident ambassadors. The Porte and the European states furthermore shared and worked with some common rules and institutions such as diplomacy, conferences, treaties and most notably Capitulations. Yet, the Porte, having intensive exchanges with Europe, was not seen within the system. The settled states in the Americas and Australasia, having less developed relations with Europe compared to the Porte, was considered within the European system. The only explanation is the cultural divide or civilizational difference.[54] Civilizational identity then influenced the development of the modern international system, especially during its European period.

One can therefore confidently conclude that some degree of civilizational identity and its consciousness in respect of the Europeans has been in operation when they dealt with non-European societies. As I have shown, the European states pursued policies towards the non-European societies not just out of their national interests or the requirements of expediency, but also out of their common identity or what was supposed to be culturally/civilizationally shared by the Europeans and not shared by the non-Europeans. Furthermore, it allows us to speak of a civilizational action or mobilization, though not to the same extent as state action or national mobilization.

The modern international system is, I have noted, both inter-civilizational and trans-civilizational. It is inter-civilizational in the sense that within the system multiple civilizational identities exist. Earlier, in Chapter 5, I have shown that there are at least five or so civilizations that still exist in the contemporary world. Because the modern international system is inter-civilizational, we do not see that the basic values, operations and rules of the system were not entirely shaped by the matrices of one single civilization, though those of Western civilization have some degree of dominance. Because it is trans-civilizational, we see a decline in the influence of civilizations and a trend towards cosmopolitanism.

On the one hand there is, it is argued, the decline of what is called 'cultural unity' of the system[55] compared to the systems that were uni-civilizational. On the other hand, because it is trans-civilizational and cosmopolitan, so it is said, there is a trend towards a common 'world-culture' or 'culture of modernity' or cosmopolitan civilization. In the words of one student, 'The "world-culture" which is emerging as the common element of middle-class life-styles in Washington and Moscow, Havana and Caracas, Dakar and Tokyo, may indeed be a rather superficial matter of parallel tastes and snob values' in a variety of things such as cars, films, music, Coca-Cola and McDonald's. 'But that does not necessarily reduce its importance for the diffusion of fellow-feeling.'[56]

I have earlier argued that a unity or wholeness of the world in empirical/historical terms could be shown and possible (Chapter 3). In line with Hodgson and McNeill's analyses, it has already been recorded that commonalities and similarities have been more and more frequent since at least AD 1000. Then, the modern international system can be seen, perhaps, as the culmination of these similarities and commonalities. Today, we see exclusive standards of 'civilization' no more. Gong makes the point that the old standard withered away because all countries are recognized as civilized, at least, according to the old standard. There is no need to say that it does not mean the end of the idea of a standard. Gong records the would-be standards as the standards of human rights and of modernity.[57] What is striking in these new candidates, we note, is that they appear to have a more universal appeal than the old ones.

Before concluding, it should, however, be recorded that the growth of unity is limited within the system due to its very nature. First, the nation-state, though transcending the civilizational boundaries, created its own limits. In the words of Suganami, 'the growth in the sense of community among mankind is itself to some extent hindered by the division of mankind into sovereign states which tend to reinforce national parochialism'.[58] Second, despite the cosmopolitan and trans-civilizational trends, civilizational identities and values have not disappeared. For ill or good, all the talk about 'the clash of civilizations' in the post-cold war era is a clear testimony to the impact of civilizational identity. Even if we accept the emerging signs of a world-culture, the multi-culturalism of the system persists. Civilizational values still have their influences.

Conclusion

The starting point of this book was the obvious uncertainty and immediate certainty of human life that takes place within a social process. Based upon this apparent feature of the social life of men, the book has examined the theoretical activity of human beings and their experience in space and time, that is to say, history. It has begun with such a discussion not only for the reason that man's history and knowledge are closely interdependent processes, but also because any theoretical or intellectual activity, such as the one undertaken in the present study, if it is to be conducted and understood clearly, requires some idea about the nature of the theoretical activity itself, as one's understanding of the theoretical activity and historical process in general shapes and directs one's analysis of the particular subject in hand.

Accordingly, I have made the point that all human knowledge, like anything human, is social and historical for it is where we, human beings, begin and end. In the human world, there is no Archimedean point on which we can rely and by which we may know once and for all. Man's sociality makes his knowledge social as well. It has been argued that objectivism and subjectivism cannot be maintained in theory, science and history. Historicalness or historicity defies objectivism as it shows that nothing is universal in the sense of being independent of time and space, and socialness or sociality defies subjectivism as it shows that no human being is self-sufficient and thus not everything goes. Knowledge is social and historical.

Being historical, there is a limit to man and his knowledge. Being social, the limits of man and his knowledge are not confined to only one category or environment. In other words, man and his knowledge are multi-faceted. The multi-faceted nature of man, his knowledge of society, his ability to abstract and his historical accumulation enable us to define various levels and units. This framework, far from constraining man's possible groupings, shows the multiplicity of his identifications and thus units. Yet, it does indicate that it is not endless. Upon this framework, I have argued, a historical conception of universal or world history is possible.

Throughout this study the historicity and continuity of human phenomena, perhaps coming from the obvious uncertainty and the immediate certainty of human life, has been taken together with its sociality. Since Aristotle, we know that, no matter on what level it is expressed and with whatever unit it is associated, the human being is a social being. Man thus leads to men. Men have various collective identifications and all collective identifications are social and historical. The historicalness of a collective identity refers, on the one hand, to the fact that it takes place within the limits of time and space and, on the other hand, to the fact that the extent and degree of its collectivity, cohesion and impact upon individual human persons may vary in time. For instance, as already stated, religion as a form of group identity was effective in medieval Europe, whereas nationality has been the dominant form in the modern period. The sociality and historicality of what are called innate identifications allow individuals to transcend, to a certain degree, those very identifications. Innate identifications are not absolute. Yet, the sociality makes their transcendence difficult, as it brings about continuity.

Of the historical, and one could say logically possible, collective identifications or units, the present study has examined the two, namely civilizations and international systems. Civilizations, as units are, as I defined, large-scale social identifications or large-scale collectivities in terms of spatial and temporal extensions. They are relatively long-lived and self-sufficient intelligible collectivities in comparison to other collectivities. Being large-scale, civilizations have a multiplicity of other social identities. Historically, most civilizations have never been embraced within one organized social entity. This is why civilizations present a low degree of cohesion and potency and are less likely to be grounds for collective mobilization. It is observed that the impact of civilizational identity upon identities and actions of individuals is not institutionalized to the degree of being exclusive in contrast to the case of political or organized social identifications.

International systems are units that form a framework in which at least two distinct human groupings, say nations, coexist. In other words, an international system is a response to the question of the peaceful coexistence of multiple units. It could be an anarchical framework in which we see loosely associated units with each having some degree of self-government, or a hierarchical framework where one unit imposes its rule over the others. In both cases, however, the problem of the coexistence of multiple units remains. Historically, just as we have no proof of one man living alone, as expressed in Aristotle's dictum 'man is a political animal', we are not in a position to prove that human beings have ever been grouped under one entity or socio-political grouping, or one society has ever lived in isolation from all the others. Expressed in another way, it is better to talk not of a single man but of men, and not of a single society but of societies. In this sense, the problem of the international system goes far back to the beginning of human society just like the problem of politics goes far back to the beginning of man.

Civilizations and international systems, both generally consisting of multiple units, are closely related to each other and historically have gone hand in hand. Most of the civilizations of which we have historical record have comprised some kind of international system. Yet, international systems could be multi-civilizational due to the looseness of the boundaries of civilizations on the one hand, and to civilizational exchanges and encounters on the other. In contrast to the idea that civilizations are self-sufficient isolated entities, incapable of understanding and interacting with each other as, for instance, Spengler assumed, civilizational encounters and exchanges may lead to international systems which are multi-civilizational as, for example, Toynbee, Wight and Bull argued. I have noted that the interactions between Western, Islamic, Hindu, Chinese and Japanese civilizations have been the central axis of world history in the last millennium. The culmination of this process has been what I have called the modern international system.

The principal and defining characteristics of the modern international system have been described in terms of an anarchical structure: centralized, territorial and sovereign nation-states being the principal form of political organization, and nationality as the main focus of social identity; humanism, secularism and rationalism as opposed to divine regulation; individualism and democracy, and industrialism. Although most of these characteristics originated from the values of Western civilization, they have universal messages. This, what Gong termed the distinction between 'universal' or 'technical' aspects of Western civilization and those aspects of strictly European or Western character, is one reason why the modern international system has become globalized. Yet, the global character of the modern international system is not just an expression of the culmination or manifestation of an idea as some, for instance Fukuyama, assumed. It is rather an expression of the unity of world achieved in the historical process in which we see some ideas and values acquiring worldwide acceptance as a result of civilizational encounters. To understand it, a historical conception of universal history is required. Such a conception, though leading to a universal or worldwide unit, is the obvious indication of the existence of different civilizational elements in the modern international system. Furthermore, if one concurs with McNeill that the Europeans had an agreement to disagree from the eighteenth century onwards, then the modern international system, though Western-dominant, can be said to embody civilizational coexistence.

The modern international system, as I have argued, has been, and is, both inter-civilizational and trans-civilizational. Because it is inter-civilizational, different civilizational elements and values have not been eliminated and they might be still effective. On the other hand, because it is trans-civilizational, the effects of civilizational elements have been lessened. Furthermore, since it has been basically Western-centred, elements and values of one civilization have prevailed. However, the nation-state as the principal focus of social identity in the modern period can hardly be said to have eliminated the civilizational identity, as the multi-faceted nature of social phenomena does not allow one unit to monopolize human social identification. As the nation-state does not encapsulate all social identifications and the predominance of Western civilization did not eliminate all other civilizations, civilizational elements and different civilizational values have existed in the modern international system.

A conception of cultural unity, such as the one assumed by Wight,[1] seems hardly possible in a multi-civilizational system. Yet, for an international system, a high degree of cultural unity is not required. Gong, though he admits Bull's view that a universal international society has come into existence in the twentieth century, notes that Japan has always had (and still has) difficulties in conforming to a standard of 'civilization'.[2] Yet, no one thinks that Japan is outside the modern international system. Of course, one can argue for the cultural unity of the contemporary international system. However, it seems to me that a high degree of cultural unity for a global international system can be argued only on the basis of what I have defined as an objectivist conception of universal history as, for example, argued by Fukuyama.

The prevalence of the nation-state and the dominance of Western civilization have, as already stated, suppressed the civilizational differences in the modern international system. Accordingly, the cold war in the second half of the twentieth century disguised the particular regional conflicts and divisions behind the ubiquitous preeminence of the East–West confrontation. It also disguised the civilizational conflicts or reduced them to a conflict between, say, civilization and barbarism in the view of both sides. One may argue that 'civilization is for most people pretty far down on the scale of self-conscious identities'.[3] One may equally argue that the end of the cold war has led to a greater awareness of civilizational elements as significant factors within the modern international system as the existing and emerging countries within and around the European hinterland are struggling to form new identities or to fortify old ones. Today we see a considerable number of peoples and movements that put emphasis on civilizational identity such as Islamic fundamentalism and European integration. We also witness a greater number of individuals in many societies stressing the importance of once-forgotten civilizational identities.

The civilizational element in world politics and the modern international system was prompted, rather provocatively, by Huntington in the last decade.[4] Arguing that after the cold war the fundamental source of conflict in the New World will be primarily cultural rather than ideological and economic and that the conflict will occur along the lines of civilizations, Huntington describes the basic reasons for and factors in 'the clash of civilizations' as follows: first, civilizational differences are far more fundamental than differences among political ideologies and political regimes; second, as the world becomes smaller the interactions between peoples of different civilizations are increasing and thus enhancing civilization-consciousness and awareness of differences between civilizations and commonalities within civilizations; third, as a result of economic modernization and social change, local identities are getting weaker and religious identity is strengthening, providing a basis for civilizational identity; fourth, the confrontation between the powerful West and non-Westerners leads the latter to return to their roots; fifth, cultural characteristics are less mutable and less easily compromised and resolved than political and economic ones; finally, economic regionalism is increasing, hence reinforcing civilization-consciousness.[5]

These observations no doubt contain some truth, though perhaps not entirely in the way Huntington envisages. Nor is he the first scholar emphasizing the role of civilizational differences. Toynbee, Wight and writers of the English School have always been conscious of the fact that the modern international system, originally emerging within a civilization, comprised civilizational differences. Gong predicted that the realm of cultural sovereignty might become the next major arena in the struggle for sphere of influence.[6] Nor could the impact of religion, especially

Islam, a major component of civilizational identity, be said to have come before the eyes of the students of international relations after the cold war. From the late 1970s onwards we see talk about the 'revival of Islam'.[7]

Those studies, however, largely considered religion as a matter of influence on policies, issues and respective governments. When we come to the 1990s, before Huntington, some began to talk of a civilizational conflict, representing a tendency to shift from international relations to inter-civilizational relations. According to Bernard Lewis, Muslim rage over the West in general and Islamic fundamentalism in particular reflect nothing less than 'a clash of civilizations'.[8] Gilpin sees the signs of civilizational conflict not only between the West and Islam, but among others as well. 'Today the revival of Islamic, Chinese and Hindu civilizations, as well as the emergence of potentially powerful new or previously isolated civilizations, in particular Japan, Brazil, and Mexico, suggest that a new era is opening.'[9] While Huntington notes trends in 'Asianization' in Japan, 'Hinduization' in India, 're-Islamization' in the Middle East and 'Russianization' in Russia, Gong observes a resurgent culturalism in Indonesia.[10] It seems fair to speak of a return to cultural or civilizational roots in many countries.

Of all the 'revivals' that have been said to be taking place in the contemporary world, the revival of Islam has been the most widely discussed. And this is not without understandable reason. It could be said that Islam, as a religion, is the most politically oriented one, for its founder, Mohammed, was also the ruler of a political community. There is no supposed distinction in Islam between the things of God and the things of Caesar, no separation of the church from the state. Moreover, the idea of a single Islamic polity transcending particular states and nations has always been very strong in Islamic history, though, except perhaps during the first centuries of Islamic history, it has never been realized. On the other hand, historically, Islam and Christianity, and thus Western civilization and Islamic civilization, have always been strong rivals and in fierce opposition. These characteristic differences and historical competition do not only account for much of the talk about Islamic revival or Islamic fundamentalism that we witness nowadays, but also for the recent events which are described as the examples of Islamic revival, from the Iranian revolution, through the victory of Islamic parties in the Algerian popular elections, to the rise of pro-Islamic parties in Turkey. According to Lewis, the fact that radical and popular movements, inaccurately called 'fundamentalist', have won mass support in Muslim countries demonstrates that the ideal of a single Islamic polity still has considerable appeal for Muslims. Islam provides the most effective form of consensus in and among Muslim countries and constitutes the basic group identity among the masses. 'Islam is', argues Lewis, 'a powerful but still undirected force in politics.'[11]

That there is an increasing identification of the masses and some elites in Muslim countries with a wider entity which is basically defined by Islam and Islamic civilization, and that a degree of collective rallying on the basis of religious (civilizational) identity in Muslim countries is possible, have been starkly demonstrated by the Gulf War of 1991 and the Bosnian War after the disintegration of Yugoslavia. According to Misha Glenny, the conflict in Bosnia 'assimilated the characteristics of a religious struggle, defined by three great European faiths – Roman Catholicism, Eastern Orthodoxy and Islam, the confessional detritus of the empires whose frontiers collided in Bosnia'.[12] As is well-known, even Iran, the fiercest enemy of the secularist regime of Iraq, gave indirect support to Iraq in the name of Islam and against the United States. In Turkey, the majority of public opinion including a considerable number of the traditional Westernized elite opposed the governmental policy of providing the allied forces with facilities that were to be used against Iraq. Similarly, the American operation against the Taliban government in Afghanistan, after the horrendous attacks on the Twin Towers of the World Trade Center in New York City on 11 September 2001, has overwhelmingly been opposed by public opinion in Muslim countries, including Turkey. There is no need to extend this list, as I have already made the point that an external threat (the American-led operation in this case) affirms the identity in concern.

The process of the formation of a wider identity based upon Islamic civilization is not something limited to the historical Middle East. It has been pointed out that the process is beginning to be effective in the ex-Soviet republics of Central Asia and Transcaucasia.[13] Upon this basis, we are justified in saying that the established national identities in countries with predominantly, Muslim populations are increasingly being questioned as the sole group identification, and Islam, as a religion and a civilization, is beginning to come to the surface as a proper identity. This process can be demonstrated in Turkey, a country with a predominantly Muslim population and a strong, even Jacobin, policy of secularization and long-standing Western orientation.

The Republic of Turkey, established in 1923, was officially declared to be a secular state from early on. The new administrative and intellectual elites of Turkey defined the new state as a 'modern' state committed to the values of Western civilization. The role of Islam in the government has been totally denied and it has been relegated to a body of beliefs and rituals observed by the majority of population. Some sort of Turkish national identity, based upon the acceptance of the 'Turkishness' of everyone within the established boundaries, was envisaged. The new state renounced any interest in the Ottoman legacy and made a strong commitment to the West. In other words, a civilizational identity based on Islam, or any wider identity of the sort, has been denied. The values of Western civilization have been perceived as not just the values of *a* civilization but as those of *the* civilization of a universal character.[14] In line with these orientations, Turkey has taken part in, or applied for membership to, organizations of Western origin such as NATO, the European Council, the OSCE, the European Community/Union and so on.

The strong Western orientation has by no means been adopted by the majority of the population still conforming to the values and elements of Islamic civilization. From the 1970s onwards, groups questioning the Republican national identity and Western orientation and arguing for a more Islamic-oriented identity and policies began to emerge. Moreover, the Western orientation has increasingly been questioned by some elites, and the Ottoman legacy is no longer being overtly renounced. It has been noted that when we came to the 1990s, after the downfall of the Soviet Union, there was an effort among the Turkish elites to try to revise their group identity and enlarge the definition of Turkey's region, leading to a search for a wider identity. On the one hand, the identity is considered as Islamic and on the other hand as Turkic, so as to include Turkic republics of Central Asia.[15] It should be noted that Turkey's new orientation towards Central Asian states cannot solely be attributed to common linguistic and ethnic characteristics, but the shared religion, Islam, as well. For example, Tajikistan, which is not Turkic but Persian-speaking, was invited to the Turkic summit held in Ankara, in October 1992.

The search for new identities or orientations in Turkey has made its impact upon the policies pursued by Turkish statesmen. Turkish policy toward the Bosnian War was a striking example. Not only did Turkish public opinion almost unanimously support the Bosnians, but Turkish governmental policies were very often at odds with its Western allies. Such a policy was not seen in the 1950s and 1960s. The Turkish government, for example, sided with France in the issue of Algerian independence. Both Bosnians and Algerians are Muslims and both countries are ex-Ottoman territories, but neither of them are Turks nor contain

Turkish-speaking peoples. I think this is due to the increasing influence of Turkey's historically held identity of Islam and Islamic civilization or due to the decreasing role of Western orientation, which was the basis of the Republican national identity.

These brief accounts of Islam in general and Turkey in particular could be taken as the concrete examples of, on the one hand, the existence of civilizational differences and, on the other hand, the effect of civilizational elements. However, upon these observations, it can hardly be argued that a full civilizational clash is on the way to replace political, ideological and economic conflicts as Huntington predicts. First of all, civilizational differences have always been there in the modern international system for it is, as stated, a multi-civilizational system. Yet, the differences in civilizations are not as acute as some doom theorists argue. Even the two historical rivals, Islam and Christendom have much in common compared to their differences. Perhaps that is why they are in dispute very often. Second, civilizations, and not least those movements arguing for an Islamic unity, are not organized entities as nations or other social identities so as to make an inter-civilizational conflict permanent and worldwide. There is neither a unified Islamic bloc nor a Western bloc. I have already made the point that civilizations being large-scale entities are not tightly organized like small identity groups. Third, the process of globalization, the shrinking of the world and worldwide communication networks, make the world a smaller place. Huntington assumes that it reinforces civilizational differences between distinct civilizations and commonalities within each civilization. It could do. Yet, it could do just the opposite as well. Many argue that a cosmopolitan international society comprising different elements and features from different cultures is possible and in the process of being created.[16] Finally, as I have already argued, civilizations, just like nation-states could coexist. One should not underestimate the adaptability of cultures and civilizations. The modern international system is a clear indication of this.

In the closing decades of the twentieth century and in the beginning of the twenty-first century, we may note two paradoxical tendencies among and within societies – localism and unionism. On the one hand, we see human beings increasingly identifying themselves with small units, whether it is an established unit like the nation-state or small local communities. Despite all the talk about the demise of the nation-state and world interdependence, people do not seem to give up the state. Even those who struggle against an existing nation-state do so in order to create another nation-state. The disintegration of the Soviet Union and former Yugoslavia increased the number of nation-states. The state and national identity built upon it, though striven for by many, has been under attack because of its restrictiveness and exclusiveness. We are only too familiar with the demands for a greater say in their own affairs by regions, cities, even neighbourhoods. In short, it would not be wrong to assert that there is a general trend of localism. That the most ambitious and perhaps the most accomplished integration movement of the century has adopted a form of 'union' (that is, the European Union) instead of its previous form of 'community' (that is, the European Community) could be taken as an indication of the appeal of 'localism'. Today the nation-state, though being a particularistic identity compared to its predecessor, does not seem to have the hold it used to have on more particularistic units of identity.

On the other hand, although we note a trend of localism and efforts to form small identity units, paradoxically we observe an orientation towards wider and larger unions with which people equally wish to identify themselves. It is true that some large states have collapsed and some others are being questioned. It is, however, equally true that the existing international and wider units still survive and there is an increasing demand upon them. The authority of 'Brussels' has been decreasing as stated, but there is a long queue at the door of 'Brussels'. Those who want to have a smaller identity want to have it together with a greater one. The Scottish nationalists aim at establishing an independent state of Scotland, a smaller identity than the United Kingdom, yet they think of not just an independent Scotland, but an independent Scotland within Europe, a greater identity than the United Kingdom. An independent Scotland is envisaged within the European Union, not within the African Union or within the Commonwealth of Independent States. Absurd as this may seem, alas it shows the effects of what may conveniently be called civilizational identity. Of course, the trend towards greater and general unions and associations can also be accounted for by the process of globalization, the shrinking of the world, worldwide networks of communication, increasing world interdependence in economy and global problems of poverty, environment, population and so on.

In spite of all this talk about globalization, why do I speak of civilizations? Surely there is no *a priori* and enduring case for any social unit of identification. However, any identification is socio-historical. I have already made the point that the socio-historicality of men is multifaceted. In other words, people are capable of holding several different identities. They could at the same time be Scottish, British, European and Western. In this study I have shown that one of these layers of identity is civilization which has been in association with the international system and it still has relevance not just as a group feeling but as a unit for international relations. The basic conclusion of this study may be expressed as the need for the modern international system, or a would-be post-modern international system, to constitute a framework in which interactions between multiple units of states, nations, societies, communities, unions and civilizations could be mediated and different values of national, social, ethnical, cosmopolitan and civilizational character can be articulated.

Notes

Introduction

1. Graham Greene, *The End of the Affair* (Harmondsworth: Penguin Books, 1962), 7.
2. The association of time with the universe and our actual experience of it being confined to the present instant had been known to the Ancient societies. Though Saint Augustine may be considered as a prominent advocate of this association and the immediacy, the Sumerians and Greeks were conscious of it. See, for instance, N. K. Sanders, *The Epic of Gilgamesh* (Harmondsworth: Penguin Books, 1960), 104; Lucretius, *The Nature of the Universe*, trans. R. E. Latham (Harmondsworth: Penguin Books, 1951), 40–1. For Saint Augustine, see Saint Augustine, *The City of God*, 2 vols, trans. John Healey, ed by R. V. G. Tasker (London: J. M. Dent, 1945), xi. 5,6; xii. 16. For a comprehensive treatment of time and its conceptions, see G. J. Whitrow, *What Is Time?* (London: Thames & Hudson, 1972) and *Time in History: Views of Time from Prehistory to the Present Day* (Oxford: Oxford University Press, 1989).
3. By 'International Relations' or 'International Theory', I mean the intellectual enterprise which mainly developed in the Anglo-Saxon world after the First World War. Principal concepts, issues, terms, subjects, techniques and methods debated by and in the academic communities which are formed by those who define themselves as the scholars of international relations can be taken as constituting the discipline of International Relations. For examples of the account of the nature and the development of the discipline of International Relations, see W. C. Olson, 'The Growth of a Discipline' in Brian Porter (ed.), *The Aberystwyth Papers: International Politics 1919–1969* (London: Oxford University Press, 1972); H. Bull, 'The Theory of International Politics 1919–1969' in B. Porter (ed.), *The Aberystwyth Papers*; S. Hoffmann, 'An American Social Science: International Relations', *Daedalus*, vol. 106, no. 3 (1977), 41–60; M. Banks, 'The Evolution of International Relations Theory' in M. Banks (ed.), *Conflict in World Society: A New Perspective on International Relations* (Brighton: Wheatsheaf Books, 1984); K. J. Holsti, *The Dividing Discipline: Hegemony and Diversity in International Relations* (London: George Allen & Unwin, 1985); S. Smith, 'Paradigm Dominance in International Relations: The Development of International Relations as a Social Science', *Millennium: Journal of International Studies*, vol. 16, no. 2 (1987), 189–206; Chris Brown, *Understanding International Relations* (Basingstoke: Macmillan – now Palgrave Macmillan, 1997); T. Dunne, M. Cox and K. Booth (eds), *The Eighty Years' Crisis: International Relations 1919–1999* (Cambridge: Cambridge University Press, 1999); and Scott Burchill et al., *Theories of International Relations*, 2nd edn (Basingstoke: Palgrave – now Palgrave Macmillan, 2001). For a historiographical account of International Relations in the United States and Britain, see, respectively, Brian C. Schmidt, *The Political Discourse of Anarchy: A Disciplinary History of International Relations* (Albany: State University of New York Press, 1998); and Timothy Dunne, *Inventing International Society: A History of the English School* (Basingstoke: Macmillan – now Palgrave Macmillan, 1998).

1 Theory or Coffee without Sugar

1. Arnold J. Toynbee, 'My View of History' in his *Civilization on Trial* (London: Oxford University Press, 1948), 3.
2. Arnold J. Toynbee, *An Historian's Approach to Religion* (London: Oxford University Press, 1956), 1.
3. Among the definitions of theory given by the *Oxford English Dictionary* are 'mental view, contemplation; a conception or mental scheme of something ..., a scheme or system of ideas; abstract knowledge'.
4. See, for instance, René Descartes, 'Discourse on the Method' in *The Philosophical Writings of Descartes*, trans. J. Cottingham et al. (Cambridge: Cambridge University Press, 1984), vol. I, 111, 127; and 'Principles of Philosophy' in *The Philosophical Writings of Descartes*, 179, 195.
5. M. Oakeshott, 'Rational Conduct' in *Rationalism in Politics and Other Essays* (London: Methuen, 1962), 90.
6. R. G. Collingwood, *The Idea of History* (Oxford: Oxford University Press, 1961), 186, 306.
7. Collingwood, *The Idea of History*, 226.
8. Collingwood, *The Idea of History*, 226, 292.
9. M. Oakeshott, 'Rational Conduct', 89–90.
10. See, C. G. Hempel, *Aspects of Scientific Explanation* (New York: Free Press, 1965), 142, 426; and Karl R. Popper, *Logic of Scientific Discovery* (New York: Basic Books, 1959), 44.
11. See, respectively, Plato, *Republic*, trans. A. D. Lindsay (London: Heron Books, in arrangement with J. M. Dent, date of publication not given), bk 7; B. Russell, *History of Western Philosophy and Its Connections with Political and Social Circumstances from the Earliest Times to the Present Day* (London: Readers Union with Allen & Unwin, 1954), 141; R. C. Zaehner, *Dawn and Twilight of Zoroastrianism* (London: Weidenfeld & Nicolson, 1961), 55; I. Kant, *Critique of Pure Reason*, trans. N. K. Smith (London: Macmillan, 1968), 163–4; and F. Nietzsche, *Twilight of the Idols*, trans. R. J. Hollingdale (London: Penguin Books, 1968), 46.
12. Hempel, *Aspects of Scientific Explanation*, 3; E. Nagel, *The Structure of Science* (London: Routledge & Kegan Paul, 1961), 80–5.
13. R. Bernstein, *Beyond Objectivism and Relativism* (Oxford: Basil Blackwell, 1983), 8.
14. Isaiah Berlin, 'History and Theory: The Concept of Scientific History', *History and Theory*, vol. 1 (1961), 29.
15. F. Nietzsche, *On the Genealogy of Morals*, trans. W. Kauffmann and R. J. Hollingdale (New York: Vintage Books, 1969), 3rd Essay, par. 12.
16. W. H. White, 'The Politics of Historical Interpretation: Discipline and De-Sublimation', *Critical Inquiry*, vol. 9 (1982), 114.
17. H. G. Gadamer, *Truth and Method* (New York: Seabury Press, 1975), 239.
18. P. G. Winch, *The Idea of a Social Science and Its Relation to Philosophy* (London: Routledge & Kegan Paul, 1958), 33–9.
19. Michael Donelan, *Elements of International Political Theory* (Oxford: Clarendon Press, 1990), 3.
20. Bernstein, *Beyond Objectivism and Relativism*, 8.
21. Russell, *History of Western Philosophy*, 98.

22. Paul K. Feyerabend, *Against Method: Outline of An Anarchistic Theory of Knowledge* (London: New Left Books, 1975), 28.
23. I owe this point to Berlin. See Isaiah Berlin, 'Introduction' in his *Four Essays on Liberty* (Oxford: Oxford University Press, 1969).
24. Berlin, 'Introduction', xxxi–xxxii, lii–liii.
25. Richard Shapcott, 'Conversation and Coexistence: Gadamer and the Interpretation of International Society', *Millennium: Journal of International Studies*, vol. 23, no. 1 (1994), 74.
26. Kuhn's paradigmatic view for science was developed in his seminal work, *The Structure of Scientific Revolutions* (Chicago: University of Chicago Press, 1st edn 1962 and 2nd edn 1970). The second edition contains a 'postscript 1969'. He further elucidated his views in his *The Essential Tension: Selected Studies in Scientific Tradition and Change* (Chicago and London: University of Chicago Press, 1977). The following analysis basically draws upon *The Structure of Scientific Revolutions* and all the references are to the second edition (1970).
27. The first one of the definitions of science given by *Oxford English Dictionary* is 'the state or fact of knowing; knowledge or cognizance of something specified or implied'.
28. T. S. Kuhn, 'The Relations between History and History of Science', *Daedalus*, vol. 100, no. 2 (1971), 271–304.
29. See, respectively, Nagel, *The Structure of Science*, 12, 448, 450, 459, 466, 473, 485; Winch, *The Idea of a Social Science*, 72, 119, 94; K. R. Popper, *The Poverty of Historicism* (London: Ark Paperbacks, 1986), 157; and A. Giddens, *Studies in Social and Political Theory* (London: Hutchinson, 1977), 26–7.
30. See H. Bull, 'International Theory: The Case for a Classical Approach', *World Politics*, vol. 18, no. 3 (1966), 361–77 and M. A. Kaplan, 'The New Great Debate: Traditionalism vs. Science in International Politics', *World Politics*, vol. 19, no. 1 (1966), 1–20. A collection of articles on 'the debate' that largely took place in the United States was provided in K. Knorr and J. N. Rosenau (eds), *Contending Approaches to International Politics* (Princeton: Princeton University Press, 1969).
31. For the effect of Kuhn on social sciences, see B. Barnes, *T. S. Kuhn and Social Science* (London: Macmillan – now Palgrave Macmillan, 1982). The implications of Kuhn's work and other 'post-positivist' philosophies of science for social disciplines have been succinctly examined in Bernstein, *Beyond Objectivism and Relativism;* and A. Giddens, *Social Theory and Modern Sociology* (Cambridge: Polity Press, 1987).
32. Kuhn, *The Structure of Scientific Revolutions*, 210.
33. Kuhn, *The Structure of Scientific Revolutions*, 176.
34. Kuhn, *The Structure of Scientific Revolutions*, 176–7.
35. Kuhn, *The Structure of Scientific Revolutions*, 175.
36. Kuhn, *The Structure of Scientific Revolutions*, 46.
37. Kuhn, *The Structure of Scientific Revolutions*, 10.
38. Kuhn, *The Structure of Scientific Revolutions*, 34–7. Such a characterization of scientific activity implies that theories of verification and falsification do not stand, because theories and tests proceed from within one or another paradigm-based tradition. Restricted in this way, the community would have no access to all possible experiences to verify the theory and they would not

choose problems that are not likely to have solutions. Furthermore, if any failure to fit in with the theory were to be grounds for theory rejection, argues Kuhn, all theories ought to be rejected at all times. See Kuhn, *The Structure of Scientific Revolutions*, 145, 147. Cf. Popper, *Logic of Scientific Discovery*, 27–30, 40.

39. Kuhn, *The Structure of Scientific Revolutions*, 52–3, 77, 82.
40. Kuhn, *The Structure of Scientific Revolutions*, 84–5, 92–4, 157–8.
41. Kuhn, *The Structure of Scientific Revolutions*, 4.
42. For criticism of Kuhn on his relativism see D. Shapere, 'The Structure of Scientific Revolutions', *Philosophical Review*, vol. 73 (1964), 383–94; K. R. Popper, 'Normal Science and Its Dangers' in I. Lakatos and A. Musgrave (eds), *Criticism and the Growth of Scientific Knowledge* (Cambridge: Cambridge University Press, 1970); and R. Keat and J. Urry, *Social Theory as Science* (London: Routledge & Kegan Paul, 1975).
43. Kuhn, *The Structure of Scientific Revolutions*, 205–7.
44. Kuhn, *The Essential Tension*, 209, 272.
45. Kuhn, *The Structure of Scientific Revolutions*, 160–1. It should be stated here that the Kuhnian and post-Kuhnian conceptions of science have given some consolation to the social scientists who are concerned about the scientificity of their subjects. See, for example, Giddens, *Social Theory and Modern Sociology*, 18, 70; M. Hesse, *Revolutions and Reconstructions in the Philosophy of Science* (Brighton: Harvester Press, 1980), 171–2; E. Gellner, 'The Scientific Status of the Social Sciences', *International Social Science Journal*, vol. 36, no. 102 (1984), 567.
46. Based on Collingwood, *The Idea of History*, 4–5.

2 History or did Napoleon Win at Waterloo?

1. See, for instance, R. Aron, 'Philosophy of History', *Chambers Encyclopedia*, vol. 7 (1946), 50; Isaiah Berlin, 'History and Theory: The Concept of Scientific History', *History and Theory*, vol. 1 (1961), 1; W. B. Gallie, *Philosophy and Historical Understanding* (London: Chatto & Windus, 1964), 51; Patrick Gardiner, *The Nature of Historical Explanation* (London: Oxford University Press, 1961), ix; A. Marwick, *The Nature of History* (London: Macmillan, 1970), 14–15; and M. Oakeshott, *On History and Other Essays* (Oxford: Basil Blackwell, 1983), 1–2.
2. See R. G. Collingwood, *An Essay on Metaphysics* (Oxford: Clarendon Press, 1940), ch. vii; and *The Idea of History* (Oxford: Oxford University Press, 1961), 226, 229, 248.
3. See F. Nietzsche, *Will to Power*, trans. W. Kauffmann and R. J. Hollingdale (London: Weidenfeld & Nicolson, 1968), par. 606; *Twilight of the Idols*, trans. R. J. Hollingdale (London: Penguin Books, 1968), 60; and *Ecce Homo: How One Becomes What One Is*, trans., R. J. Hollingdale (London: Penguin Books, 1969), 70.
4. K. Mannheim, *Ideology and Utopia* (London: Routledge & Kegan Paul, 1936), 71.
5. Michael Donelan, 'Introduction' in M. Donelan (ed.), *The Reason of States: A Study in International Political Theory* (London: G. Allen & Unwin, 1978), 18, 22; and 'The Political Theorists and International Theory' in Donelan (ed.), *The Reason of States*, 85–6.

6. G. W. F. Hegel, *Lectures on the Philosophy of World History* ed by Johannes Hoffmeister, trans. H. B. Nisbet (Cambridge: Cambridge University Press, 1975), 29.
7. Collingwood, *The Idea of History*, 248, 282.
8. Ibn Khaldun, *The Muqaddimah: An Introduction to History*, trans. F. Rosenthal, 2nd edn (London: Routledge & Kegan Paul, 1967), vol. I, 6, 71, 77–8.
9. See, respectively, G. Vico, *The New Science*, trans. T. B. Gergen and M. H. Fisch (Ithaca: Cornell University Press, 1948), bk. 2, 100; I. Kant, 'Idea of a Universal History from a Cosmopolitan Point of View' in Patrick Gardiner (ed.), *Theories of History* (New York: Free Press, 1959), 22; Hegel, *Lectures on the Philosophy of World History*, 44; Arnold J. Toynbee, *A Study of History* (London: Oxford University Press, 1934), vol. 3, 446–8; and Collingwood, *The Idea of History*, 9, 211–16, 302.
10. See J. G. von Herder, *Reflections on the Philosophy of* the *History of Mankind*, abridged and with an introduction by F. E. Manuel (Chicago and London: University of Chicago Press, 1968); and R. Aron, *Introduction to the Philosophy of History: An Essay on the Limits of Historical Objectivity*, trans. G. J. Irwin (London: Weidenfeld & Nicolson, 1961), 15.
11. Collingwood, *The Idea of History*, 216, 308–9.
12. Fernand Braudel, 'History and Social Sciences: The Long Term', *Social Science Information*, vol. 9, no. 1 (1970), 155.
13. Immanuel Wallerstein, 'World System Analysis' in A. Giddens and J. Turner (eds), *Social Theory Today* (Cambridge: Polity Press, 1987), 324.
14. Sir John Seeley, *Introduction to Political Science* (London: Macmillan, 1896), 4.
15. The literature is vast. See, for instance, Collingwood, *The Idea of History*, 5–9; Gallie, *Philosophy and Historical Understanding*, 51–2; F. Guizot, *Historical Essays and Lectures*, ed and with an introduction by S. Mellon (Chicago and London: University of Chicago Press, 1972), 7; Oakeshott, *On History and Other Essays*, 1–2, 27; Leopold von Ranke, *The Secret of World History: Selected Writings on the Art and Science of History*, ed with translations by Roger Wines (New York: Fordham University Press, 1981), 58; and Toynbee, *A Study of History*, vol. I, 446–8.
16. See H. Butterfield, *The Whig Interpretation of History* (New York: Norton, 1965), 47; Oakeshott, *On History and Other Essays*, 9; J. Ortega y Gasset, *An Interpretation of Universal History*, trans. M. Adams (New York and London: W. W. Norton, 1973), 49; L. von Ranke, 'On the Relation of and Distinction between History and Politics' in his *The Secret of World History*, 114.
17. Braudel, 'History and Social Sciences ...', 154; Edmund Burke, *Reflections on the Revolution in France* (London: J. M. Dent, 1910), 93. F. Nietzsche, *Thus Spake Zarathustra*, trans. A. Tille (London: J. M. Dent, 1933), 126.
18. For Greene, see the Introduction and for Oakeshott, see *On History and Other Essays*, 8. See also G. J. Whitrow, *Time in History: Views of Time from Prehistory to the Present Day* (Oxford: Oxford University Press, 1989), 5.
19. Collingwood, *The Idea of History*, 10.
20. R. G. Collingwood, 'The Three Laws of Politics' *Hobhouse Memorial Lectures, 1941–1950* (London: Oxford University Press, 1952), 26.
21. Kant, 'Idea of a Universal History ...', 22; Hegel, *Lectures on the Philosophy of World History*, 28; and K. Marx and F. Engels, 'Manifesto of the

Communist Party' in K. Marx and F. Engels, *Selected Works* (London: Lawrence & Wishart, 1968).

22. Collingwood, *The Idea of History*, 224.
23. W. H. Walsh, *An Introduction to Philosophy of History*, 3rd edn (London: Hutchinson, 1967), 13.
24. Gallie, *Philosophy and Historical Understanding*, 55.
25. Ranke, *Secret of World History*, 58.
26. L. von Ranke, 'A Dialogue on Politics' in T. H. von Laue (ed.), *Leopold Ranke: The Formative Years* (Princeton: Princeton University Press, 1950), 165.
27. Quoted in 'Introduction' to Leopold von Ranke, *Secret of World History*, 21.
28. Von Laue, *Leopold Ranke: The Formative Years*, 43.
29. See, for instance, A. Sked, 'The Study of International Relations: A Historian's View', *Millennium: Journal of International Studies*, vol. 16, no. 2 (1987), 251–62. For a thorough discussion of historians and objectivity, see also Peter Novick, *That Noble Dream: The 'Objectivity Question' and the American Historical Profession* (New York: Cambridge University Press, 1988); and John Lewis Gaddis, 'History, Theory and Common Ground', *International Security*, vol. 22, no. 1 (Summer 1997), 75–85.
30. H. T. Buckle, *Introduction to the History of Civilization in England*, revd edn by J. M. Robertson (London: George Routledge, date of publication not given), 3–4, 18.
31. Seeley, *Introduction to Political Science*, 6, 29.
32. I. Berlin, 'Historical Inevitability' in his *Four Essays on Liberty* (Oxford: Oxford University Press, 1969), 57–8.
33. J. B. Bury, 'The Science of History' in *Selected Essays of J. B. Bury* ed. by H. Temperley (Cambridge: Cambridge University Press, 1930), 19, 22.
34. Collingwood, *The Idea of History*, 9, 249.
35. See, respectively, G. Ritter, 'Scientific History, Contemporary History, and Political Science', *History and Theory*, vol. 1 (1961), 267; and J. M. Clubb, 'History as a Social Science', *International Social Science Journal*, vol. 33, no. 4 (1981), 596.
36. See, for example, W. Dray, *Laws and Explanation in History* (London: Oxford University Press, 1957), 44; and Walsh, *An Introduction to Philosophy of History*, 39.
37. See Berlin, 'History and Theory ...', 8–9, 22–3, 30; C. B. Joynt and N. Rescher, 'The Problem of Uniqueness in History', *History and Theory*, vol. 1 (1961), 152–3; and E. H. Carr, *What is History?*, 2nd edn by R. W. Davies (London: Penguin Books, 1987), 69.
38. The allusion is to Graham Greene's *The Human Factor* (London: Penguin Books, 1978), which could be taken as showing the violating impact of human beings even in the case of a strict control mechanism within a long-standing organization.
39. F. Meinecke, 'Values and Causalities in History' in F. Stern (ed.), *The Varieties of History: From Voltaire to the Present* (Cleveland: Meridian Books, 1956), 268–88; and C. A. Beard, 'That Noble Dream' in F. Stern (ed.), *The Varieties of History*, 315–28. On 'new historicism', see, for instance, H. A. Veeser (ed.), *The New Historicism* (New York and London: Routledge, 1989); Marjorie Levinson, et al., *Rethinking Historicism* (Oxford: Basil Blackwell, 1989). On the 'new philosophy of history' and post-modernism, see, for example, Leonard

Krieger, *Time's Reasons: Philosophies of History Old and New* (Chicago and London: University of Chicago Press, 1989); Frank R. Ankersmit and Hans Kellner (eds), *A New Philosophy of History* (Chicago: University of Chicago Press, 1995); Lynn Hunt and Jacques Revel (eds), *Histories: French Constructions of the Past* (New York: New Press, 1995); and Keith Jenkins (ed.), *Postmodern History Reader* (New York and London: Routledge, 1997).

40. Meinecke, 'Values and Causalities in History,' 268, 273; and Beard, 'That Noble Dream', 316, 325–8.
41. Paul Valéry, *History and Politics*, trans. D. Folliot and J. Mathews (London: Routledge & Kegan Paul, 1962), 511, 522.
42. Note that this conception of historicism is different from Popper's definition which means 'an approach to the social sciences which assumes that *historical prediction* is their principal aim, and which assumes that this aim is attainable by discovering the "rhythms" or the "patterns", the "laws" or the "trends" that underlie the evolution of history.' K. R. Popper, *The Poverty of Historicism* (London: Ark Paperbacks, 1986), 3.
43. L. Strauss, 'Political Philosophy and History', *Journal of the History of Ideas*, vol. 10, no. 1 (1949), 46.
44. See, for instance, Hans Kellner, *Language and Historical Representation: Getting the Story Crooked* (Madison: University of Wisconsin Press, 1989), 4; John H. Zammito, 'Are We Being Theoretical Yet? The New Historicism, The New Philosophy of History and "Practicing Historians" ', *Journal of Modern History*, vol. 65, no. 4 (1993), 145–6; and F. Ankersmit, 'Reply to Professor Zagorin', *History and Theory*, vol. 29 (1990), 281.
45. See, for example, Lionel Gossman, *Towards a Rational Historiography, Transactions of the American Philosophical Society*, vol. 79, pt 3 (Philadelphia: American Philosophical Society, 1989), 51, 54; Stephen Bann, *The Invention of History: Essays on the Representation of the Past* (Manchester: Manchester University Press, 1990), 41; Zammito, 'Are We Being Theoretical Yet? ...', 812; and Perez Zagorin, 'Historiography and Postmodernism: Reconsiderations', *History and Theory*, vol. 29 (1990), 266.
46. L. Wittgenstein, *Philosophical Investigations*, trans. G. E. M. Anscombe (Oxford: Basil Blackwell, 1978), par. 43; R. G. Collingwood, *The New Leviathan or Man, Society, Civilization and Barbarism*, revd edn, ed and introduced by D. Boucher (Oxford: Clarendon Press, 1992), xvii; and *The Principles of Art* (London: Oxford University Press, 1938), 250.
47. For instance, the view that Kuhn is relevant to all fields of history finds its expression in David A. Hollinger, 'T. S. Kuhn's Theory of Science and Its Implications for History', *American Historical Review*, vol. 78, (1973), 370–93; see also his 'The Return of the Prodigal: The Persistence of Historical Knowledge', *American Historical Review*, vol. 94 (1989), 610–21; and Randolph Starn, 'Historians and "Crisis" ', *Past and Present*, no. 52 (August 1971), 3–22. For Isaac Kramnick, Kuhn makes a significant and original contribution to the theories of social change. See his 'Reflections on Revolution: Definitions and Explanation in Recent Scholarship', *History and Theory*, vol. 11 (1972), 26–63. Skinner attempted to apply a 'set of concepts' to the history of ideas, similar to that applied to the history of science by Kuhn. See Quentin Skinner, 'Meaning and Understanding in the History of Ideas', *History and Theory*, vol. 8 (1969), 3–53.

48. J. Burckhardt, *Reflections on History*, trans. by M. D. H. (London: George Allen & Unwin, 1943), 74.
49. Collingwood, *The Idea of History*, 248.
50. O. Spengler, *The Decline of the West*, trans. C. F. Atkinson (New York: Alfred A. Knopf, 1926), Vol. I, 60; Toynbee, *Civilization on Trial*, 3–4; and G. Barraclough, *History in a Changing* World (Oxford: Basil Blackwell, 1957), 14.
51. Hollinger, 'T. S. Kuhn's Theory of Science and Its Implications for History', 382.
52. T. S. Kuhn, 'Reflections on My Critics' in I. Lakatos and A. Musgrave (eds), *Criticism and the Growth of Knowledge* (Cambridge: Cambridge University Press, 1970), 244–5.
53. A. J. P. Taylor, *The Origins of the Second World War* (London: Hamish Hamilton, 1961). For the critics, see W. M. R. Louis (ed.), *The Origins of the Second World War: A. J. P. Taylor and Its Critics* (New York: Wiley, 1972).
54. A. J. Toynbee, *Hannibal's Legacy: The Hannibalic War's Effects on Roman Life*, 2 vols (London: Oxford University Press, 1965); *A Study of History*, 12 vols (London: Oxford University Press, 1934–61). For reviews of *Hannibal's Legacy*, see 'Toynbee Returns to Rome', *Times Literary Supplement* (December 1965); *Classical Philology* vol. 62, no. 2 (April 1967), 144–6, by S. I. Oost; *American Historical Review*, vol. 72, no. 2 (January 1967), 537–9, by T. R. S. Broughton; *The New Statesman and Nation*, vol. 70, no. 1815 (24 December 1965), 1003–4, by M. I. Finley; and *History*, vol. 51, no. 172 (June 1966), 199–201, by H. H. Scullard. For criticisms of *A Study of History*, see M. F. Ashley Montagu (ed.), *Toynbee and History: Critical Essays and Reviews* (Boston: Porter Sargent Publisher, 1956).
55. H. G. Wells, *The Outline of History: Being a Plain History of Life and Mankind* (London: Cassell, 1920). For a criticism, see G. Barraclough, 'Universal History' in H. P. R. Finberg (ed.), *Approaches to History: A Symposium* (London: Routledge & Kegan Paul, 1962).

3 Universal History or the World as we Know It

1. For an examination of the problem of the 'unit of analysis', see A. Nuri Yurdusev, ' "Level of Analysis" and "Unit of Analysis": A Case for Distinction', *Millennium: Journal of International Studies*, vol. 22, no. 1, (1993), 77–88.
2. Herbert Butterfield, *Man on His Past: The Study of the History of Historical Scholarship* (Cambridge: Cambridge University Press, 1955), 45.
3. Butterfield, *Man on His Past*, 49–50.
4. W. H. McNeill, *A World History* (London: Oxford University Press, 1967), 24.
5. For the inception of the three-fold division, see Butterfield, *Man on His Past*, 45–6. For critics, see G. Barraclough, *History in a Changing World* (Oxford: Basil Blackwell, 1957), 57, 231; A. J. Toynbee, *A Study of History* (London: Oxford University Press, 1934), vol. I, 170–1; and O. Spengler, *The Decline of the West*, trans. C. F. Atkinson (New York: Alfred A. Knopf, 1926), vol. I, 16–18.
6. Saint Augustine, *The City of God*, 2 vols, trans. John Healey, ed by R. V. G. Tasker, with an Introduction by Sir Ernest Barker, Everyman edition (London: J. M. Dent, 1945).
7. R. G. Collingwood, *The Idea of History* (Oxford: Oxford University Press, 1961), 32–3.

8. N. Berdyaev, *The Meaning of History*, trans. G. Reavey (London: Geoffrey Bles, The Centenary Press, 1936), 27–33. For the contrary view, see Hans Kohn, *The Idea of Nationalism: A Study in Its Origins and Background* (New York: The Macmillan Company, 1944), 34.
9. A. R. Coll, *The Wisdom of Statecraft: Sir Herbert Butterfield and the Philosophy of International Politics* (Durham: Duke University Press, 1985), 22.
10. H. Butterfield, *The Origins of History* (London: Eyre Methuen, 1981), 88.
11. Quoted in Butterfield, *Man on His Past*, 123.
12. H. Butterfield, *Christianity and History* (London: G. Bell, 1949), 3.
13. Berdyaev, *The Meaning of History*, 123.
14. Saint Augustine, *The City of God*, xii.
15. C. Dawson, *The Dynamics of World History* (London: Sheed & Ward, 1957), 235–6.
16. Dawson, *The Dynamics of World History*, 315.
17. St. Augustine, *The City of God*, vol. II, 408.
18. A. Linklater, *Men and Citizens in the Theory of International Relations* (London: Macmillan – now Palgrave Macmillan, 1982), 48. Linklater provides a good analysis of the rationalist universalism of the seventeenth and eighteenth centuries in part two.
19. It is not easy to distinguish these terms – progress, development and evolution, as all have very similar connotations. For an attempt at distinction, see, for instance, M. Ginsberg, *The Idea of Progress: A Revaluation* (London: Methuen, 1933), 41–2.
20. J. B. Bury, *The Idea of Progress: An Inquiry into Its Origin and Growth* (London: Macmillan, 1928), 1–36.
21. Dawson, *The Dynamics of World History*, 246.
22. I. Kant, 'Idea of a Universal History from a Cosmopolitan Point of View' in P. Gardiner (ed.), *Theories of History* (Glencoe, IL: Free Press, 1959), 22–3.
23. Kant, 'Idea of a Universal History', 30.
24. Kant, 'Idea of a Universal History', 32–3.
25. Kant, 'Idea of a Universal History', 24–6, 28.
26. G. W. F. Hegel, *Lectures on the Philosophy of World History* ed by Johannes Hoffmeister, trans. H. B. Nisbet (Cambridge: Cambridge University Press, 1975), 128.
27. Hegel, *Lectures on the Philosophy of World History*, 27–9, 39.
28. See, respectively, L. von Ranke, 'Introduction' in his *The Secret of World History: Selected Writings on the Art and Science of History*, ed., with translation by Roger Wines (New York: Fordham University Press, 1981), 25. Emphasis added; and 'The Great Powers' in his *The Secret of World History*, 154; J. Burckhardt, *Reflections on History*, trans. M. D. H. (London: George Allen & Unwin, 1943), 22–3; A. J. Toynbee, *A Study of History* (London: Oxford University Press, 1954), vol. IX, 395.
29. See, for instance, J. B. Noone, Jr., 'The Philosophy of History: A Prolegomenon to Political Philosophy', *The Review of Politics*, vol. 23 no. 4 (1961), 481.
30. See, J. G. von Herder, *Reflections on the Philosophy of the History of Mankind*, abridged and with an introduction by F. E. Manuel (Chicago and London: University of Chicago Press, 1968), 80; H. Butterfield, *The Whig Interpretation of History* (New York: Norton, 1965), 65; and *Christianity and History*, 29–30;

W. H. Walsh, *An Introduction to Philosophy of History*, 3rd edn (London: Hutchinson, 1967), 63–5.

31. Kant, 'Idea of a Universal History', 22, 24; Hegel, *Lectures on the Philosophy of World History*, 30.
32. T. H. von Laue, *Leopold Ranke: The Formative Years* (Princeton: Princeton University Press, 1950), 122–5.
33. I. Berlin, 'Historical Inevitability' in his *Four Essays on Liberty* (Oxford: Oxford University Press, 1969), 60.
34. What Ranke alludes to is the publication of the record of different nations under the title of *An Universal History from the Earliest Account of Time to the Present Compiled from the Original Authors*, 38 vols (London, 1763–1765). See Butterfield, *Man On His Past*, 47, 123–4.
35. Ranke, 'A Dialogue on Politics' in von Laue, *Leopold Ranke: The Formative Years*, 164–5; and L. von Ranke, *Universal History: The Oldest Historical Groups of Nations and the Greeks*, trans. and ed by G. W. Prothero (London: Kegan Paul, Trench, 1884), xi–xiv.
36. J. Acton, *Lectures on Modern History* ed with an introduction by J. N. Figgis and R. V. Laurence (London: Macmillan, 1950), 317. The idea of the unity of history and the whole is also expressed by, for instance, the following authors: Burckhardt, *Reflections on History*, 22; J. B. Bury, 'The Science of History' in *Selected Essays of J. B. Bury*, ed by H. Temperley (Cambridge: Cambridge University Press, 1930), 8; Toynbee, *A Study of History*, vol. I, 22; H. G. Wells, *The Outline of History: Being a Plain History of Life and Mankind* (London: Cassell, 1920), 2; J. Ortega y Gasset, *An Interpretation of Universal History*, trans. M. Adams (New York and London: W. W. Norton, 1973), 40; G. Barraclough, 'The Larger View of History', *Times Literary Supplement*, no. 2810 (6 January 1956), ii; *History in a Changing World*, 29; and L. S. Stavrianos, 'The Teaching of World History', *The Journal of Modern History*, vol. 31 (1959), 115.
37. See, respectively, Kant, Idea of a Universal History', 25; Hegel, *Lectures on the Philosophy of World History*, 65, 93, 96; and K. Marx and F. Engels, 'Manifesto of the Communist Party' in K. Marx and F. Engels, *Selected Works* (London: Lawrence & Wishart, 1968), 52.
38. Cf. Butterfield, *Christianity and History*, 66–7.
39. Barraclough, *History in a Changing World*, 15.
40. Barraclough, *History in a Changing World*, 29.
41. A. Kuzminski, 'Archetypes and Paradigms: History, Politics, and Persons', *History and Theory*, vol. 25, no. 3 (1986), 227.
42. Butterfield, *Man on His Past*, 109–10, 113.
43. See Collingwood, *The Idea of History*, 90.
44. Ranke quoted in A. F. Wright, 'The Study of Chinese Civilization', *Journal of the History of Ideas*, vol. 21 (1960), 245.
45. Barraclough, *History in a Changing World*, 15,17.
46. See Barraclough, *History in a Changing World* and G. Barraclough, 'Universal History' in H. P. R. Finberg (ed.), *Approaches to History: A Symposium* (London: Routledge & Kegan Paul, 1962); and McNeill, *A World History* and 'Organizing Concepts for World History', *Review*, vol. 10, no. 2 (1986), 211–29.
47. Marshall G. S. Hodgson (ed.), *Rethinking World History: Essays on Europe, Islam and World History*, with an Introduction and Conclusion by E. Burke, III (Cambridge: Cambridge University Press, 1993).

48. Perez Zagorin, 'Historiography and Postmodernism: Reconsiderations', *History and Theory*, vol. 29 (1990), 263–74.
49. Toynbee, *A Study of History*, vol. I, 17–22.
50. Ranke, 'The Great Powers', 122.
51. Ranke, 'A Dialogue on Politics', 165.
52. Ortega y Gasset, *An Interpretation of Universal History*, 16.
53. L. von Ranke, 'History and Philosophy' in *The Secret of World History*, 103.
54. McNeill, *A World History*, 19–42; and 'Organizing Concepts for World History', 225. For a similar view, see Marshall G. S. Hodgson, 'The Interrelations of Societies in History' in his *Rethinking World History: Essays on Europe, Islam and World History*, ed., with an Introduction and Conclusion by E. Burke, III (Cambridge: Cambridge University Press, 1993).
55. L. von Ranke, 'On the Relation of and Distinction between History and Politics' in *The Secret of World History*, 115; and 'The Great Powers', 155.
56. Barraclough, *History in a Changing World*, 19, 29.
57. E. A. Freeman, 'Historical Cycles' in his *Historical Essays*, fourth series (London: Macmillan, 1892), 257.
58. Barraclough, 'Universal History', 100–1; and 'The Larger View of History', ii.
59. Barraclough, 'Universal History', 102.
60. Stavrianos, 'The Teaching of World History', 115. See also Barraclough, 'Universal History', 83.
61. C. K. Webster, 'The Study of International History', *History*, vol. 17 (1933), 99–100.
62. P. Savigear, 'International Relations and the Philosophy of History' in M. Donelan (ed.), *The Reason of States* (London: George Allen & Unwin, 1978), 203.

4 Civilization or Naked Greed

1. For Ibn Khaldun's use of the word *umran*, see 'Translator's Introduction' in Ibn Khaldun, *The Muqaddimah: An Introduction to History*, trans., with Introduction, F. Rosenthal, 2nd edn (London: Routledge & Kegan Paul, 1967), xxvi. For the meaning of the Chinese word *wen*, see A. F. Wright, 'The Study of Chinese Civilization', *Journal of the History of Ideas*, vol. 21, (1960), 234–5.
2. This summary of the early usages of the word civilization is based upon Fernand Braudel, 'The History of Civilizations: The Past Explains the Present' in his *On History*, trans. S. Matthews (London: Weidenfeld & Nicolson, 1980), 180; *A History of Civilizations*, trans. Richard Mayne (Harmondsworth: Penguin Books, 1993), 3–8; Norbert Elias, *The History of Manners*, trans. E. Jephcott (New York: Pantheon Books, 1978), 38; A. L. Kroeber and C. Kluckhohn, *Culture: A Critical Review of Concepts and Definitions* (Cambridge, Mass.: Pleabody Museum of American Archeology and Ethnology, 1952), 1, 37; and *Oxford English Dictionary* (1989).
3. Elias, *The History of Manners*, 53–4.
4. Braudel, 'The History of Civilizations', 180; and *A History of Civilizations*, 5.
5. Kroeber and Kluckhohn, *Culture*, 43–72.
6. R. H. Lowie, *The History of Ethnological Theory* (New York: Rinehart and Company, 1937), 10.

7. See, for instance, G. P. Gooch, *History and Historians in the Nineteenth Century*, 2nd edn (London: Longmans, 1952), 523.
8. F. Guizot, *The History of Civilization in Europe* (London: Cassell, 1911), 5–8, 11–15.
9. J. S. Mill, 'Civilization', *The London and Westminster Review*, vol. 28, no. 1 (1836), 1, 3–6, 14–17.
10. J. Arthur de Gobineau, *The Inequality of Human Races*, trans. A. Collins (London: William Heinemann, 1915), 91.
11. K. Marx and F. Engels, 'Manifesto of the Communist Party' in K. Marx and F. Engels, *Selected Works* (London: Lawrence & Wishart, 1968), 40.
12. F. Engels, 'Origins of Family, Private Property and State' in Marx and Engels, *Selected Works*, 582.
13. Braudel, 'The History of Civilizations', 181–2; and *A History of Civilizations*, 5–6.
14. Elias, *The History of Manners*, 4.
15. O. Spengler, *The Decline of the West*, 2 vols, trans. C. F. Atkinson (New York: Alfred A. Knopf, vol. I 1926 and vol. II 1928), vol. I, 31,106.
16. E. B. Tylor, *Primitive Culture: Researches into the Development of Mythology, Philosophy, Religion, Language, Art, and Custom* (New York: Brentanos Publishers, 1903), 1.
17. Ibn Khaldun, *The Muqaddimah*, vol. I, 89; vol. II, 270–1, 291.
18. Braudel, 'The History of Civilizations', 177; and Elias, *The History of Manners*, 5.
19. M. Melko, *The Nature of Civilizations* (Boston: Porter Sargent Publisher, 1969), 8.
20. M. G. S. Hodgson, *The Venture of Islam: Conscience and History in a World Civilization*, 3 vols (Chicago: The University of Chicago Press, 1971), vol. I, 91.
21. P. Bagby, *Culture and History: Prolegomena to the Comparative Study of Civilizations* (London: Longmans, 1958), 84, 162–3.
22. See, for instance, Kroeber and Kluckhohn, *Culture*, 13–17; and T. S. Eliot, *Notes Towards the Definition of Culture* (London: Faber & Faber, 1948), 13.
23. Elias, *The History of Manners*, 3.
24. Cf. Aristotle, *Politics*, trans. J. Warrington (London: Heron Books in arrangement with J. M Dent, date of publication not given), bk 1, 5–8; Ibn Khaldun, *The Muqaddimah*, vol. I, 89; vol. II, 270–1; and R. G. Collingwood, *The New Leviathan or Man, Society, Civilization and Barbarism*, revd edn by D. Boucher (Oxford: Clarendon Press, 1992), 283.
25. Collingwood, *The New Leviathan*, 283–4.
26. Guizot, *The History of Civilization in Europe*, 6.
27. Braudel, 'The History of Civilizations', 180–1; and *A History of Civilizations*, 6–7. For Herder, see J. G. Herder, *Reflections on the Philosophy of the History of Mankind* (Chicago and London: University of Chicago Press, 1968), 159.
28. Lowie, *The History of Ethnological Theory*, 13, 20, 59.
29. See, especially, Engels, 'Origins of Family, Private Property and State'.
30. Tylor, *Primitive Culture*, 26.
31. See W. K. C. Guthrie, *A History of Greek Philosophy* (Cambridge: Cambridge University Press, 1969), vol. 3, 82.
32. Edward Gibbon, *The History of the Decline and Fall of the Roman Empire* ed. by J. B. Bury (London: Methuen, 1925), vol. IV, 167.
33. Lowie, *The History of Ethnological Theory*, 20.
34. Sir John Lubbock, *The Origin of Civilization and the Primitive Condition of Man* (London: Longmans, Green, 1911), 384.

35. Tylor, *Primitive Culture*, 29, 31. Despite this judgement, Tylor earlier qualified the primitive condition as a 'hypothetical state'. *Primitive Culture*, 21.
36. See S. N. Kramer, *The Sumerians* (Chicago and London: University of Chicago Press, 1963), 262.
37. See, respectively, *The New English Bible: With the Apocrypha* (London: Oxford University Press and Cambridge University Press, 1970), 2; and *The Koran: Interpreted*, trans., with an Introduction, A. J. Arberry (Oxford: Oxford University Press, 1983), 5.
38. See, for example, F. Boas, 'The Limitations of the Comparative Method of Anthropology', *Science*, vol. 4 (18 December 1896), 901–8.
39. Tylor, *Primitive Culture*, 7.
40. Lowie, *The History of Ethnological Theory*, 59.
41. Tylor, *Primitive Culture*, 26–7.
42. See, respectively, W. Eckhardt, *Civilizations, Empires and Wars: A Quantitative History of War* (Jefferson, N.C. and London: McFarland, 1992), 7; Q. Wright, *A Study of War*, 2nd edn (Chicago: Chicago University Press, 1965), 55.
43. Bagby, *Culture and History*, 184.
44. V. G. Childe, *Man Makes Himself* (London: Watts, 1936); and *What Happened in History* (Harmondsworth: Penguin Books, 1942).
45. Engels, 'Origins of Family, Private Property and State', 572.
46. R. Redfield, *The Primitive World and Its Transformations* (Ithaca: Cornell University Press, 1953), 30; and H. Frankfort, *The Birth of Civilization in the Near East* (London: Williams & Norgate, 1951), 57–8.
47. Bagby, *Culture and History*, 162–3, 184.
48. W. H. McNeill, 'Organizing Concepts for World History', *Review*, vol. 10, no. 2 (1986), 221; and 'Civilization', 3.
49. W. H. McNeill, *A World History* (London: Oxford University Press, 1967), 13; and *The Global Condition: Conquerors, Catastrophes, and Community* (Princeton: Princeton University Press, 1992), 81–2.
50. A. J. Toynbee, *A Study of History* (London: Oxford University Press, 1934), vol. I, 118–204; vol. III, 242.
51. H. N. Spalding, *Civilization in East and West: An Introduction to the Study of Human Progress* (London: Oxford University Press, 1930), 2–3.
52. Lord Raglan, *How Came Civilization* (London: Methuen, 1939), 3, 176–7; and Eliot, *Notes Towards the Definition of Culture*, 15, 28, 31.
53. Mill, 'Civilization', 2–3.
54. C. Quigley, *The Evolution of Civilizations: An Introduction to Historical Analysis* (New York: The Macmillan Company, 1961), 31.
55. Collingwood, *The New Leviathan*, 487.
56. Collingwood, *The New Leviathan*, 489.

5 Civilizations or Realities of the Extreme *Longue Dureé*

1. C. Levi-Strauss, *Structural Anthropology* (London: Macmillan, 1976), 35; J. Derrida, *The Other Heading: Reflections on Today's Europe*, trans. P. Brault and M. B. Naas (Bloomington: Indiana University Press, 1992), 9–10.
2. Thucydides, *History of the Peloponnesian War*, trans. R. Crawley (London: J. M. Dent, 1910), 3.

3. G. M. Trevelyan, *History of England* (London: Longmans, 1943), 232.
4. See, for instance, Paul Coles, *The Ottoman Impact on Europe* (London: Thames & Hudson, 1968); I. B. Neumann and J. M. Welsh, 'The Other in European Self-Definition: An Addendum to the Literature on International Society', *Review of International Studies*, vol. 17 (1991), 327–48; I. B. Neumann, *The Uses of the Other: The East in European Identity Formation* (Manchester: Manchester University Press, 1999); and A. Nuri Yurdusev, 'Turkey and Europe: The Other in Identity Formation', *Zeitschrift fur Turkeistudien*, 13. Jahrgang 2000, Heft 1, 85–94.
5. W. Bloom, *Personal Identity, National Identity and International Relations* (Cambridge: Cambridge University Press, 1990), 23.
6. A. D. Smith, *National Identity* (Harmondsworth: Penguin Books, 1991), 175.
7. Bloom, *Personal Identity*, 32, 39.
8. The understanding of a collective identification in terms of objective and subjective elements is a practical and functional account that is quite common in the literature. See, for instance, Smith on nation and *ethnié, National Identity*; Kohn on nation and nationalism, Hans Kohn, *The Idea of Nationalism: A Study in Its Origins and Background* (New York: The Macmillan Company, 1944); Huntington on civilization, S. P. Huntington, 'The Clash of Civilizations?', *Foreign Affairs*, vol. 72, no. 3 (Summer 1993), 22–49 and *The Clash of Civilizations and the Remaking of the World Order* (New York: Simon Schuster, 1996); Bull on 'society of states', Hedley Bull, *The Anarchical Society: A Study of Order in World Politics* (London: Macmillan, 1977); and Vincent on 'race', R. J. Vincent, 'Race in International Relations', *International Affairs*, vol. 58, no. 4 (1982), 658–70.
9. Denys Hay, *Europe: Emergence of an Idea*, 2nd edn (Edinburgh: Edinburgh University Press, 1968); and Maurice Keens-Soper, 'The Practice of a States System' in M. Donelan (ed.), *The Reason of States: A Study in International Political Theory* (London: G. Allen & Unwin, 1978).
10. Smith, *National Identity*, 4–7, 9–14, 21.
11. Kohn, *The Idea of Nationalism*, 5–6.
12. Bloom, *Personal Identity*, 74.
13. Fernand Braudel, 'The History of Civilizations: The Past Explains the Present' in his *On History*, trans. S. Matthews (London: Weidenfeld & Nicolson, 1980), 209–10; and *A History of Civilizations*, trans. R. Mayne (Harmondsworth: Penguin Books, 1993), 34–5.
14. Arnold J. Toynbee, *A Study of History*, 3 vols (London: Oxford University Press, 1934), vol. I, 45.
15. Marshall G. S. Hodgson, *The Venture of Islam: Conscience and History in a World Civilization*, 3 vols (Chicago: The University of Chicago Press, 1974), vol. I, 91; and William H. McNeill, 'The Changing Shape of World History', *History and Theory, Theme Issue 34, World Historians and Their Critics*, vol. 34, no. 2 (1995), 13.
16. S. P. Huntington, 'The Clash of Civilizations?', 24; and *The Clash of Civilizations*, 43.
17. O. Spengler, *The Decline of the West*, 2 vols, trans. C. F. Atkinson (New York: Alfred A. Knopf, vol. I 1926 and vol. II 1928), vol. I, 21–2, 59–60, 104.
18. Toynbee, *A Study of History*, vol. I, 22–6, 455.
19. See, L. von Ranke, his Introduction to the *History of the Latin and Teutonic Nations*, in *The Secret of World History* (New York: Fordham University Press,

1981), 56; and R. G. Collingwood, *The Idea of History* (Oxford: Oxford University Press, 1961), 14.

20. See Marshall, G. S. Hodgson, 'The Interrelations of Societies in History' in his *Rethinking World History: Essays on Europe, Islam, and World History*, ed., with an Introduction and Conclusion by E. Burke III (Cambridge: Cambridge University Press, 1993); 1–28: and McNeill, 'The Changing Shape of World History', 13.
21. J. Arthur de Gobineau, *The Inequality of Human Races*, trans. A. Collins (London: William Heinemann, 1915), 212.
22. Quoted in A. Linklater, *Men and Citizens in the Theory of International Relations* (London: Macmillan – now Palgrave Macmillan, 1982), 21–2.
23. See Herodotus, *The Persian Wars* (Harmondsworth: Penguin Books, 1972), 543–4.
24. A. L. Kroeber, 'The Delimitation of Civilizations', *Journal of the History of Ideas*, vol. 14, no. 2 (1953), 269–70.
25. F. S. C. Northrop, *The Meeting of East and West: An Inquiry Concerning World Understanding* (New York: The Macmillan Company, 1946), 401.
26. See, respectively, Toynbee, *A Study of History*, vol. I, 34–5; C. Dawson, *The Dynamics of World History* (London: Sheed & Ward, 1957), 128; Huntington, *The Clash of Civilizations*, 42; and T. S. Eliot, *Notes Towards the Definition of Culture* (London: Faber & Faber, 1948), 15, 28, 31.
27. Braudel, *A History of Civilizations*, 9.
28. W. Eckhardt, *Civilizations, Empires and Wars: A Quantitative History of War* (Jefferson, N.C. and London: McFarland, 1992), 43.
29. A. L. Kroeber, *Styles and Civilizations* (Ithaca: Cornell University Press, 1957), 4–6, 19–20.
30. Kroeber, 'The Delimitation of Civilizations', 273–5.
31. W. H. McNeill, 'Civilization', *The Encyclopedia Americana*, international edn (Danbury: Grolier Incorporated, 1982), 2.
32. Eliot speaks of the existence of a unified, distinctive European literature. See T. S. Eliot, 'The Unity of European Culture', in his *Notes Towards the Definition of Culture*, 112.
33. R. W. Southern, *The Making of the Middle Ages* (London: Hutchinson University Library, 1953), 50.
34. See, respectively, Hodgson, *The Venture of Islam*, vol. I, 91; McNeill, 'The Changing Shape of World History', 16.
35. Kroeber, *Styles and Civilizations*, 134.
36. W. H. McNeill, *A World History* (London: Oxford University Press, 1967), 19–42; and 'Organizing Concepts for World History', *Review*, vol. 10, no. 2 (1986), 223–5.
37. Toynbee, *A Study of History*, vol. I, 45; and P. A. Sorokin, *Modern Historical and Social Philosophies* (New York: Dover Publications, 1963), 49.
38. See de Gobineau, *The Inequality of Human Races*, 210–12. For Danilevsky's list, see Sorokin, *Modern Historical and Social Philosophies*, 49–120.
39. Spengler, *The Decline of the West*, vol. I, 18; and vol. II, 38–46, 189–96.
40. Toynbee, *A Study of History*, vol. I, 133; vol. II, 322–4; and vol. III, 1–79.
41. Hodgson, *Rethinking World History*, 8.
42. McNeill, *A World History*, 1–4.
43. Huntington, 'The Clash of Civilizations?', 25; and *The Clash of Civilizations*, 45–7.

44. McNeill, 'Organizing Concepts for World History', 225.
45. Braudel, 'The History of Civilizations', 201; Eckhardt, *Civilizations, Empires and Wars*, 35; and M. Melko, *The Nature of Civilizations* (Boston: Porter Sargent Publisher, 1969), 17.
46. Melko, *The Nature of Civilizations*, 15–16.
47. G. W. Gong, *The Standard of 'Civilization' in International Society* (Oxford: Clarendon Press, 1984).
48. Plato, *The Republic*, trans. A. D. Lindsay (London: Heron Books, in arrangement with J. M. Dent, no date), 162; and Aristotle, *Politics*, trans. and ed by J. Warrington (London: Heron Books in arrangement with J. M. Dent, no date), 6, 13, 92.
49. Gong, *The Standard of 'Civilization' in International Society*, 3–53.
50. A. F. Wright, 'The Study of Chinese Civilization', *Journal of the History of Ideas*, vol. 21, (1960), 236–7.
51. For the text of the letter, see A. F. Whyte, *China and Foreign Powers* (London: Milford, 1927). For this famous expedition of Lord Macartney to the Chinese Court, see Alain Peyrefitte, *The Collision of Two Civilizations: The British Expedition to China 1792–4*, trans. J. Rothschild (London: Harvill, 1993).
52. Toynbee, *A Study of History*, vol. II, 378.
53. McNeill, *A World History*, 416.
54. McNeill, *A World History*, 417–21.
55. McNeill, 'The Changing Shape of World History', 17.
56. The literature on international system is vast. I have extensively analysed it elsewhere. See A. Nuri Yurdusev, 'The Concept of International System as a Unit of Analysis', *METU Studies in Development* (Ankara), vol. 21, no. 1 (1994), 143–74. For a concise analysis of the concept of international system and its understandings in the United States and Great Britain, see R. Little, 'The Systems Approach' in S. Smith (ed.), *International Relations: British and American Approaches* (Oxford: Basil Blackwell, 1985).
57. M. Wight (ed.), *Systems of States*, with an Introduction by H. Bull (Leicester: Leicester University Press, 1977), 22.
58. Bull, *The Anarchical Society*, 15–16.
59. Melko, *The Nature of Civilizations*, 116.
60. F. S. Northedge, *The International Political System* (London: Faber & Faber, 1976), 37–50.
61. Adam Watson, *The Evolution of International Society: A Comparative Historical Analysis* (London: Routledge, 1992).
62. Wight, *Systems of States*, 24–5.

6 Modern International System I or No Rock without a Flag

1. Hereafter, the terms 'the modern international system', 'world-wide international system' and 'global international system' will be used interchangeably, unless, of course, stated otherwise.
2. See G. R. Berridge, *International Politics: States, Power and Conflict since 1945*, 2nd edn (London: Harvester Wheatsheaf, 1992).
3. P. Valéry, *History and Politics*, trans. D. Folliot and J. Mathews (London: Routledge & Kegan Paul, 1962), 14–15.

4. See, respectively, J. Bodin, *Six Books of the Commonwealth*, abridged and trans. by M. Y. Tooley (Oxford: Basil Blackwell, 1955), ch. x; Max Weber, 'Politics as a Vocation' in *From Max Weber: Essays in Sociology*, trans. and ed by H. H. Gerth and C. Wright Mills (New York: Oxford University Press, 1946), 77–128; and F. H. Hinsley, *Sovereignty*, 2nd edn (Cambridge: Cambridge University Press, 1986), 1, 26.
5. K. Marx and F. Engels, 'Manifesto of the Communist Party' in *Selected Works* (London: Lawrence & Wishart, 1968), 37.
6. J. Hoffman, 'Hedley Bull's Conception of International Society and the Future of the State', *Leicester University Discussion Papers in Politics*, No. 92/2, (March 1992), 9–23. See also his *Beyond the State: An Introductory Critique* (Cambridge: Polity Press, 1995).
7. Some idea of the complexity of contemporary theorizations on the state can be gleaned from, say, K. Dyson, *The State Tradition in Western Europe* (Oxford: Martin Robertson, 1980); M. Carnoy, *The State and Political Theory* (Princeton: Princeton University Press, 1984); A. Giddens, *The Nation State and Violence* (Cambridge: Polity Press, 1985); A. Vincent, *Theories of the State* (Oxford: Basil Blackwell, 1987); D. Held, *Political Theory and the Modern State: Essays on State, Power and Democracy* (Cambridge: Polity Press, 1989); and Hoffman, *Beyond the State*.
8. Quoted in Elie Kedourie, *Nationalism* (London: Hutchinson, 1960), 14–15.
9. For Mill's view, see J. S. Mill, *Considerations on Representative Government* (Oxford: Basil Blackwell, 1946), 291 (the edition also contains *On Liberty*); and for Renan, see quoted in M. Donelan, *Elements of International Political Theory* (Oxford: Clarendon Press, 1990), 79.
10. A. D. Smith, *National Identity* (London: Penguin Books, 1991), 100.
11. Smith, *National Identity*, 14–15. Emphasis added. The link between state and nation has been stressed by others as well. For instance, see H. Kohn, *The Idea of Nationalism: A Study in Its Origins and Background* (New York: The Macmillan Company, 1944), 3–4; Kedourie, *Nationalism*, 12–13; A. B. Bozeman, *Politics and Culture in International History* (Princeton: Princeton University Press, 1960), 453; B. Anderson, *Imagined Communities: Reflections on the Origins and Spread of Nationalism* (London: Verso, 1983), 15.
12. See, respectively, John Herz, *International Politics in the Atomic Age* (New York: Columbia University Press, 1959) and *The Nation State and the Crisis of World Politics* (New York: David McKay, 1976); R. O. Keohane and J. S. Nye (eds), *Transnational Relations and World Politics* (Cambridge, Mass.: Harvard University Press, 1972) and *Power and Interdependence* (Boston: Litlle Brown, 1977).
13. J. Piscatori, 'Islam in International Order' in H. Bull and A. Watson (eds), *The Expansion of International Society* (Oxford: Clarendon Press, 1984), 320–1. Cf. Berridge, *International Politics*, 9.
14. J. Dunn, *Western Political Theory in the Face of the Future* (Cambridge: Cambridge University Press, 1979), 64; and P. Kennedy, *The Rise and Fall of the Great Powers: Economic Change and Military Conflict from 1500 to 2000* (London: Fontana Press, 1989), 567.
15. H. J. Morgenthau, *Politics among Nations: The Struggle for Power and Peace*, 3rd edn (New York: Alfred A. Knopf, 1960), 10.
16. Cf. F. S. Northedge, *The International Political System* (London: Faber & Faber, 1976), 36; M. Wight, *Systems of States* (Leicester: Leicester University Press,

1977), 22–4; and A. Watson, *The Evolution of International Society: A Comparative Historical Analysis* (London: Routledge, 1992), 13–18.

17. The modern nature of national identity and nation states can easily be seen in the works of such scholar as J. Acton, *Lectures on Modern History*, ed with an introduction by J. N. Figgis and R. V. Laurence (London: Macmillan, 1950), and 'Nationality' in his *Essays on Freedom and Power* (New York: Meridian Books, 1955); Kohn, *The Idea of Nationalism*; Kedourie, *Nationalism*; Smith, *National Identity*; and Anderson, *Imagined Communities*.
18. W. H. McNeill, *A World History* (London: Oxford University Press, 1967), 56, 58.
19. Quoted in G. Barraclough, *History in a Changing World* (Oxford: Basil Blackwell, 1957), 98.
20. See, respectively, J. Burckhardt, *The Civilization of the Renaissance in Italy*, trans. by S.G.C. Middlemore (London: The Phaidon Press, 1944), 120; and M. Oakeshott, 'Rationalism in Politics' in his *Rationalism in Politics and Other Essays* (London: Methuen, 1962), 1–2. Cf. Owen Chadwick, *The Secularization of European Society in the Nineteenth Century* (Cambridge: Cambridge University Press, 1975), 264.
21. See Burckhardt, *The Civilization of the Renaissance in Italy*, 81–2; and A. Bozeman, 'The International Order in a Multicultural World' in H. Bull and A. Watson (eds), *The Expansion of International Society* (Oxford: Clarendon, 1984), 390. See also C. Taylor, *Sources of the Self: Making of Modern Identity* (Cambridge, MA: Harvard University Press, 1989).
22. A. J. Lerner, 'Transcendence of the Nation: National Identity and the Terrain of the Divine', *Millennium: Journal of International Studies*, vol. 20, no. 3 (1991), 407. See also Kohn, *The Idea of Nationalism*, 192, 196, 215–18; and C. L. Becker, *The Heavenly City of the Eighteenth-Century Philosophers* (New Haven: Yale University Press, 1932).
23. See, for instance, Q. Wright, *A Study of War*, 2nd edn (Chicago: Chicago University Press, 1965), 206; and F. Engels, 'Feuerbach and End of Classical German Philosophy' in K. Marks and F. Engels, *Selected Works*, 614. In this respect, the most well-known works from students of International Relations are by Immanuel Wallerstein. See his *The Modern World-System* (New York: Academic Press, 1974) and *The Capitalist World-Economy* (Cambridge: Cambridge University Press, 1979).
24. Notable examples analysing the historical development of the modern international system as such include H. Bull, *The Anarchical Society: A Study of Order in World Politics* (London: Macmillan, 1977); H. Bull and A. Watson (eds), *The Expansion of International Society* (Oxford: Clarendon, 1984); G. W. Gong, *The Standard of 'Civilization' in International Society* (Oxford: Clarendon, 1984); McNeill, *A World History*; G. Modelski, *Long Cycles in World Politics* (London: Macmillan – now Palgrave Macmillan, 1987); Northedge, *The International Political System*; Wallerstein, *The Modern World-System*; Watson, *The Evolution of International Society*; and Wight, *Systems of States*.
25. See for example, J. Acton, *Lectures on Modern History*, 31; O. Gierke, *Political Theories of the Middle Ages*, trans. F. Maitland (Cambridge: Cambridge University Press, 1900), 19; Kohn, *The Idea of Nationalism*, 82–4; Watson, *The Evolution of International Society*, 139–140; and Wight, *Systems of States*, 26–8. One should not forget that this picture was the depiction of only what is called Western Christendom.

26. Barraclough, *History in a Changing World*, 97–130.
27. R. W. Southern, *The Making of the Middle Ages* (London: Hutchinson, 1953), 18.
28. Figgis quoted in Wight, *Systems of States*, 28.
29. See Kohn, *The Idea of Nationalism*, 84–5.
30. Kohn, *The Idea of Nationalism*, 188.
31. M. Oakeshott, 'Character of a Modern European State', *On Human Conduct* (Oxford: Clarendon Press, 1975), 185.
32. See Barraclough, *History in a Changing World*, 102; and Wight, *Systems of States*, 27, 132–3.
33. Cf. F. Guizot, *The History of Civilization in Europe* (London: Cassell, 1911), 268–9; Acton, *Lectures on Modern History*, 71 ff.; and Kohn, *The Idea of Nationalism*, 119 ff., 188.
34. Cf. Kohn, *The Idea of Nationalism*, 132–3; and McNeill, *A World History*, 297–9.
35. Watson, *The Evolution of International Society*, 170. See also Burckhardt, *The Civilization of the Renaissance in Italy*, 207; and E. A. Freeman, 'National Prosperity and the Reformation' in his *Historical Essays*, 4th series (London: Macmillan and Co., 1892), 289.
36. McNeill, *A World History*, 313.
37. Based on Bull, *The Anarchical Society*, 31; Burckhardt, *The Civilization of the Renaissance in Italy*, 61; M. Forsyth, 'The Tradition of International Law' in T. Nardin and D. R. Mapel (eds), *Traditions of International Ethics* (Cambridge: Cambridge University Press, 1992), 36; Denys Hay, *Europe: The Emergence of an Idea* (Edinburgh: Edinburgh University Press, 1957), 54–5, 87, 96, 110, 116–17, 120–2; F. H. Hinsley, *Power and the Pursuit of Peace* (Cambridge: Cambridge University Press, 1963), 156; M. Keens-Soper, 'The Practice of a States-System' in M. Donelan (ed.), *The Reason of States* (London: G. Allen & Unwin, 1978), 27–8; Northedge, *The International Political System*, 51; F. Voltaire, *The Age of Louis XIV* (London: J. M. Dent, 1935), 5; and A. J. Toynbee, *A Study of History* (London: Oxford University Press, 1954), vol. VIII, 708–29.
38. For examples of the works emphasizing the role of the Muslim, especially the Ottoman, stranglehold over the Mediterranean and the Levant, see Barraclough, *History in a Changing World*, 132–3; J. Bowle, *The Unity of European History: A Political and Cultural Survey* (London: Jonathan Cape, 1948), 175; and A. J. Toynbee, *A Study of History* (London: Oxford University Press, 1934), vol. III, 199.
39. A. Watson, 'European International Society and Its Expansion' in Bull and Watson (eds), *The Expansion of International Society*, 27–30.
40. McNeill, *A World History*, 416–17; Modelski, *Long Cycles in World Politics*, 25–6; Wallerstein, *The Modern World System*, 16–17, 38, 67–8; and *The Capitalist World Economy* (Cambridge: Cambridge University Press, 1979), 19.
41. A. J. Toynbee, *An Historian's Approach to Religion* (London: Oxford University Press, 1956), 144–5.
42. Wright, *A Study of War*, 251–2, 858, 1520.
43. M. Howard, 'The Military Factor in European Expansion' in Bull and Watson (eds), *The Expansion of International Society*, 36, 38, 40–1.
44. Gong, *The Standard of 'Civilization' in International Society*, 97–8.
45. McNeill, *A World History*, 356.
46. See Gong, *The Standard of 'Civilization' in International Society*, 97; and McNeill, *A World History*, 294–5.

47. W. H. McNeill, *The Global Condition: Conquerors, Catastrophes, and Community* (Princeton: Princeton University Press, 1992), 123–4.
48. McNeill, *A World History*, 292–3. For an analysis of the devastating effects of the price revolution upon the Ottoman economy in the sixteenth century, see Ömer Lütfi Barkan, 'The Price Revolution of the Sixteenth Century: A Turning Point in the Economic History of the Near East', *International Journal of Middle Eastern Studies*, vol. 6 (1975), 3–28.
49. Chadwick, *The Secularization of European Society*, 264.
50. Gong, *The Standard of 'Civilization' in International Society*, 4.
51. Gong, *The Standard of 'Civilization' in International Society*, 149. For the Ottoman attitude, see T. Naff, 'The Ottoman Empire and the European States System' in Bull and Watson (eds), *The Expansion of International Society*; and J. C. Hurewitz, 'Ottoman Diplomacy and the European States System', *The Middle East Journal*, vol. 15 (Spring 1961), 141–52.
52. G. Barraclough, *An Introduction to Contemporary History* (Harmondsworth: Pelican Books, 1967), 96.
53. Bozeman, *Politics and Culture in International History*, 6.; and 'The International Order in a Multicultural World' in Bull and Watson (eds), *The Expansion of International Society*, 396. Cf. Kedourie, *Nationalism*; and 'Introduction' in *Nationalism in Asia and Africa* (ed), with an introduction, by E. Kedourie (London: Frank Cass, 1970).
54. Quoted in R. J. Vincent, 'Racial Equality' in Bull and Watson (eds), *The Expansion of International Society*, 248.
55. Gong, *The Standard of 'Civilization' in International Society*, 99–100.

7 Modern International System II or the White Man's Burden

1. H. Butterfield, *Man on His Past* (Cambridge: Cambridge University Press, 1955), 129–130, 132–3.
2. L. Von Ranke, 'Introduction to the *History of the Latin and Teutonic Nations*' in his *The Secret of World History* (New York: Fordham University Press, 1981), 57.
3. John Acton, *Lectures on Modern History* (London: Macmillan, 1950), 3, 34, 49–51, 71.
4. L. Dehio, *The Precarious Balance: Four Centuries of the European Power Struggle* (New York: Alfred A. Knopf, 1962), 23; and G. Modelski, *Long Cycles in World Politics* (Basingstoke: Macmillan – now Palgrave Macmillan, 1987), 68–9.
5. F. Guizot, *The History of Civilization in Europe* (London: Cassell, 1911), 244–5.
6. See S. Leathes (ed.), *The Cambridge Modern History* (Cambridge: Cambridge University Press, 1903), 1–2.
7. J. Bowle, *The Unity of European History: A Political and Cultural Survey* (London: Jonathan Cape, 1948), 193–5.
8. W. H. McNeill, *A World History* (London: Oxford University Press, 1967), 283–7.
9. H. Bull, *The Anarchical Society* (London: Macmillan, 1977), 14; and F. S. Northedge, *The International Political System* (London: Faber & Faber, 1976), 55.
10. J. Burckhardt, *The Civilization of the Renaissance in Italy* (London: The Phaidon Press, 1944), 106.

11. Wight, *Systems of States* (Leicester: Leicester University Press, 1977), 111–12, 131, 151–3; and I. Wallerstein, *The Modern World-System* (New York: Academic Press, 1974), 68.
12. F. H. Hinsley, *Power and the Pursuit of Peace* (Cambridge: Cambridge University Press, 1963), 153–5.
13. Butterfield, *Man on His Past*, 135.
14. G. Barraclough, *History in a Changing World* (Oxford: Basil Blackwell, 1957), 61.
15. Northedge, *The International Political System*, 55, 59.
16. For a comprehensive analysis of the Treaty of Westphalia of 1648 and the Westphalian thesis, see Andreas Osiander, *The States System of Europe 1640–1990: Peacemaking and the Conditions of International Stability* (Oxford: Clarendon Press, 1994); and 'Sovereignty, International Relations, and the Westphalian Myth', *International Organization*, vol. 55 (2001), 251–87.
17. See, for example, McNeill, *A World History*, 416 ff.; Northedge, *The International Political System*, 74–6; H. Bull, 'The Emergence of a Universal International Society' in H. Bull and A. Watson (eds), *The Expansion of International Society* (Oxford: Clarendon Press, 1984), 117–26; R. J. Vincent, 'Racial Equality' in Bull and Watson (eds), *The Expansion of International Society*, 240–1; A. Watson, *The Evolution of International Society* (London: Routledge, 1992), 265–309; A. Watson, 'European International Society and Its Expansion' in Bull and Watson (eds), *The Expansion of International Society*, 27–32; and G. W. Gong, *The Standard of 'Civilization' in International Society* (Oxford: Clarendon Press, 1984), 29–30, 41–2, 85–7.
18. Wight, *Systems of States*, 151.
19. A. B. Bozeman, *Politics and Culture in International History* (Princeton: Princeton University Press, 1960), 61–2.
20. See D. Hume, 'Of the Balance of Power' in M. Wright (ed.), *Theory and Practice of the Balance of Power 1486–1914: Selected European Writings* (London: J. M. Dent, 1975); and Thucydides, *History of the Peloponnesian War* (London: J. M. Dent, 1910), 16.
21. Bozeman, *Politics and Culture in International History*, 219–20. Cf. Barraclough, *History in a Changing World*, 26; Watson, *The Evolution of International Society*, 138–51; and H. Butterfield, *The Whig Interpretation of History* (New York: Norton, 1965), 28.
22. Bull and Watson (eds), *The Expansion of International Society*, 2–3.
23. W. H. McNeill, 'The Changing Shape of World History', *History and Theory, Theme Issue 34, World Historians and Their Critics*, vol. 34, no. 2 (1995), 17–18.
24. R. W. Southern, *The Making of the Middle Ages* (London: Hutchinson University Library, 1953), 13, 32, 42–9.
25. G. Mattingly, *Renaissance Diplomacy* (Harmondsworth: Penguin Books, 1955), 136–7.
26. See H. Inalcik, 'The Turkish Impact on the Development of Modern Europe' in K. H. Karpat (ed.), *The Ottoman State and Its Place in World History* (Leiden: E. J. Brill, 1974), 51–2; T. Naff, 'The Ottoman Empire and the European States System' in H. Bull and A. Watson (eds), *The Expansion of International Society*, 145–6; Watson, 'European International Society and Its Expansion', 16–17.
27. Dehio, *The Precarious Balance*, 40–1.
28. Inalcik, 'The Turkish Impact on the Development of Modern Europe', 52–3; Mattingly, *Renaissance Diplomacy*, 169–70; Naff, 'The Ottoman Empire and

the European States System', 176–8; Watson, *The Evolution of International Society*, 177–8, 216.

29. Inalcik, 'The Turkish Impact on the Development of Modern Europe', 56–7; H. Inalcik, *The Economic and Social History of the Ottoman Empire* (Cambridge: Cambridge University Press, 1994), 188, 366; H. Inalcik, 'Imtiyazat', *Encyclopedia of Islam*, 2nd edn (Leiden: E. J. Brill, 1971), 1179; and N. Sousa, *The Capitulatory Regime of Turkey: Its History, Origin and Nature* (Baltimore: Johns Hopkins Press, 1933), 16.
30. Maxime Rodinson, *Europe and the Mystique of Islam* (London: I. B. Tauris, 1987), 34–5. See also the works by Inalcik cited in note 29.
31. H. Butterfield, 'The Balance of Power' in H. Butterfield and M. Wight (eds), *Diplomatic Investigations: Essays in the Theory of International Politics* (London: George Allen & Unwin, 1966), 143; and R. H. Davidson, 'Ottoman Diplomacy and Its Legacy' in L. Carl Brown (ed.), *Imperial Legacy: The Ottoman Imprint on the Balkans and the Middle East* (New York: Columbia University Press, 1996), 175.
32. In addition to the works already mentioned such as Inalcik, Mattingly and Dehio see also Dorothy M. Vaughan, *Europe and the Turk: A Pattern of Alliances, 1350–1700* (Liverpool: Liverpool University Press, 1954); H. Butterfield, 'The New Diplomacy and Historical Diplomacy' in Butterfield and Wight (eds), *Diplomatic Investigations*; and Aristide R. Zolberg, 'Origins of the Modern World System: A Missing Link', *World Politics*, vol. 23, no. 2 (1981), 253–81.
33. B. Buzan and R. Little, *International Systems in World History: Remaking the Study of International Relations* (Oxford: Oxford University Press, 2000), 20.
34. Marshall G. S. Hodgson, *Rethinking World History* (Cambridge: Cambridge University Press, 1993), 68.
35. The literature on the European or Western civilization is vast. Some idea about its characteristics and extensions, origins and affiliations with Christianity and the Graeco-Roman world can, for instance, be gleaned from the following: F. Guizot, *The History of Civilization in Europe* (London: Cassell & Company, 1911); A. J. Toynbee, *A Study of History*, vol. I (London: Oxford University Press, 1934); H. Kohn, *The Idea of Nationalism* (New York: The Macmillan Company, 1944); Bozeman, *Politics and Culture in International History*; Bowle, *The Unity of European History*; Barraclough, *History in a Changing World*; McNeill, *A World History*; Denys Hay, *Europe: The Emergence of an Idea* (Edinburgh: Edinburgh University Press, 1957); Watson, *The Evolution of International Society*; Valéry, *History and Politics* (London: Routledge & Kegan Paul, 1962); and Wight, *Systems of States*.
36. Rudyard Kipling, 'White Man's Burden', *McClure's Magazine*, 12 (February 1899).
37. Quoted in N. Elias, *The History of Manners*, trans. E. Jephcott (New York: Pantheon Books, 1978), 50.
38. Vincent, 'Racial Equality', 239–40.
39. See Elias, *The History of Manners*, 53; and Gong, *The Standard of 'Civilization' in International Society*, 51.
40. F. S. Northedge and M. J. Grieve, *A Hundred Years of International Relations* (London: Duckworth, 1971), 4; and K. M. Panikkar, *Asia and Western Dominance* (London: George Allen & Unwin, 1953), 494.

41. Gong, *The Standard of 'Civilization' in International Society*, 14–15, 21.
42. Gong, *The Standard of 'Civilization' in International Society*, 10, 238.
43. See Gong, *The Standard of 'Civilization' in International Society*, 5, 141, 159–63, 188; Northedge and Grieve, *A Hundred Years of International Relations*, 7–8; and Watson, 'European International Society and Its Expansion', 25. Cf. Rodney Gilbert, *The Unequal Treaties* (London: John Murray, 1929), vii; and F. C. Jones, *Extraterritoriality in Japan and the Diplomatic Relations Resulting in Its Abolition 1853–1899*, reprinted edn (New York: AMS Press, 1970), 58.
44. Naff, 'The Ottoman Empire and the European States System', 143. Emphases added.
45. For the text of the Peace Treaty (Paris) of 1856, see J. C. Hurewitz (ed.), *The Middle East and North Africa in World Politics: A Documentary Record*, 2nd edn, vol. 1 European Expansion, 1535–1914 (New Haven and London: Yale University Press, 1975). For an assessment of the Treaty see A. Oaks and R. B. Mowat, *The Great European Treaties of the Nineteenth Century* (Oxford: Oxford University Press, 1918); H. Temperley, 'The Treaty of Paris of 1856 and Its Execution', *Journal of Modern History*, vol. 4 (1932), 387–414, 523–43; H. McKinnon Wood, 'The Treaty of Paris and Turkey's Status in International Law', *American Journal of International Law* (1943), 264–74.
46. See I. B. Neumann and J. M. Welsh, 'The Other in European Self-Definition: An Addendum to the Literature on International Society', *Review of International Studies*, vol. 17 (1991), 327–48.
47. A. Hourani, *Europe and the Middle East* (Berkeley: University of California Press, 1980), 10–12.
48. B. Lewis, *The Emergence of Modern Turkey* (London: Oxford University Press, 1961), 40.
49. J. J. Rousseau, *The Social Contract and Discourses*, trans. G. D. H. Cole (London: J. M. Dent 1913), 109.
50. P. J. Marshall and G. Williams, *The Great Map of Mankind: British Perceptions of the World in the Age of the Enlightenment* (London: J. M. Dent, 1982), 165.
51. J. Lorimer, *Institutes of the Law of Nations*, 2 vols (Edinburgh and London: William Blackwood & Sons, 1883), 10–12, 102.
52. C. Elliot, *Turkey in Europe*, 2nd edn (London: Frank Cass, 1965), 16.
53. To be fair, it may be better to state here that not all authors demonized the Turks. Bodin praised the degree of religious toleration existing under the rule of the Sultan, and Lady Mary envied the degree of freedom the Turkish women had and appreciated the kindness of the Turkish men. See, Bodin quoted in P. Coles, *The Ottoman Impact on Europe* (London: Thames & Hudson, 1968), 151; and Lady Mary, *The Turkish Embassy Letters* (London: 1763) in *The Complete Letters of Lady Mary Wortley Montagu*, vol. I, ed with an Introduction by R. Halsband (London: Oxford University Press, 1965).
54. Indeed, Adam Watson conceded that he and Bull treated the issue in this way in *The Expansion of International Society*. See Adam Watson, 'Hedley Bull, States Systems and International Societies', *Review of International Studies*, vol. 13 (1987), 147–53.
55. See, for instance, Bull, *The Anarchical Society*, 38–9.
56. R. Dore, 'Unity and Diversity in World Culture' in Bull and Watson (eds), *The Expansion of International Society* (Oxford: Clarendon Press, 1984), 419. On

the culture of 'modernity', see Bull, *The Anarchical Society*, 39, 316–17; and Gong, *The Standard of 'Civilization' in International Society*, 92–3. For scepticism about the concept of a world culture, see I. Wallerstein, 'The National and Universal: Can There be such a Thing as World Culture?' in his *Geopolitics and Geoculture: Essays on the Changing World-System* (Cambridge: Cambridge University Press, 1991).

57. Gong, *The Standard of 'Civilization' in International Society*, 241–2, 91–2.
58. H. Suganami, *The Domestic Analogy and World Order Proposals* (Cambridge: Cambridge University Press, 1989), 206. Cf. R. B. J. Walker, *Inside/Outside: International Relations as Political Theory* (Cambridge: Cambridge University Press, 1993), 117.

Conclusion

1. Martin Wight, *Systems of States* (Leicester: Leicester University Press, 1977), 33–4.
2. G. W. Gong, *The Standard of 'Civilization' in International Society* (Oxford: Clarendon Press, 1984), 200. For Bull's conception of international society and his distinction between international society and international system, see Hedley Bull, *The Anarchical Society* (London: Macmillan, 1977), 9–16.
3. See Robert W. Cox, 'Thinking about Civilizations', *Review of International Studies*, vol. 26 (2000), 127.
4. See S. P. Huntington, 'The Clash of Civilizations?', *Foreign Affairs*, vol. 72, no. 3 (1993), 22–49; and *The Clash of Civilizations and the Remaking of the World Order* (New York: Simon & Schuster, 1996).
5. Huntington, 'The Clash of Civilizations?', 25–8.
6. Gong, *The Standard of 'Civilization' in International Society*, 247.
7. For early analyses of the 'revival of Islam' or what is loosely called 'political Islam', see M. Ayob (ed.), *The Politics of Islamic Reassertion* (New York: St. Martin's Press – now Palgrave Macmillan, 1981); M. Curtis (ed.), *Religion and Politics in the Middle East* (Boulder: Westview Press, 1982); A. Dawisha (ed.), *Islam in Foreign Policy* (Cambridge: Cambridge University Press, 1983); A. E. Dessouki (ed.), *Islamic Resurgence in the Arab World* (New York: Praeger, 1982); G. H. Jansen, *Militant Islam* (New York: Harper & Row, 1979); B. Lewis, 'The Return of Islam' in his *Islam and the West* (New York and Oxford: Oxford University Press, 1993) (originally published in *Commentary* (January 1976), 39–49); E. Mortimer, *Faith and Power: The Politics of Islam* (London: Faber & Faber, 1982); and D. Pipes, *In the Path of God: Islam and Political Power* (New York: Basic Books, 1983). Since the late 1970s and early 1980s, the literature has grown enormously. For an admirable treatment of the so-called 'political Islam', see John L. Esposito, *The Islamic Threat: Myth or Reality?*, 2nd edn (Oxford: Oxford University Press, 1995).
8. B. Lewis, 'The Roots of Muslim Rage', *The Atlantic Monthly* (September 1990), 60.
9. R. Gilpin, 'The Global Political System' in J. D. B. Miller and R. J. Vincent (eds), *Order and Violence: Hedley Bull and International Relations* (Oxford: Clarendon Press, 1990), 139.
10. See, respectively, Huntington, 'The Clash of Civilizations?', 26; and Gong, *The Standard of 'Civilization' in International Society*, 245.

11. Lewis, 'The Return of Islam', 5, 153–4. For an analysis of the pros and cons of Islam being a political force in contemporary international politics, see B. Beeley, 'Islam as a Global Political Force' in A. G. McGrew and P. G. Lewis et al. (eds), *Global Politics: Globalization and the Nation-State* (Cambridge: Polity Press, 1992).
12. Misha Glenny, 'Carnage in Bosnia, for Starters', *New York Times* (29 July 1993), A23.
13. See, for example, G. E. Fuller, 'The Emergence of Central Asia', *Foreign Policy*, No. 78 (Spring 1990), 49–67; and B. Lewis, 'Rethinking the Middle East', *Foreign Affairs*, vol. 71, no. 4 (1992), 99–119.
14. For a comprehensive analysis of the emergence of the modern Turkish state, see B. Lewis, *The Emergence of Modern Turkey* (London: Oxford University Press, 1961).
15. See, for example, G. E. Fuller and I. O. Lesser, with P. B. Henze and J. F. Brown, *Turkey's New Geopolitics: From the Balkans to Western China* (Boulder: Westview Press, 1993); G. Winrow, *Where East Meets West: Turkey and The Balkans* (London: Institute for European Defence and Strategic Studies, 1993); and P. Robins, 'Between Sentiment and Self-Interest: Turkey's Policy Toward Azerbaijan and the Central Asian States', *Middle East Journal*, vol. 47, no. 4 (1993), 593–610.
16. See, for instance, R. Shapcott, 'Conversation and Coexistence: Gadamer and the Interpretation of International Society', *Millennium: Journal of International Studies*, vol. 23, no. 1 (1994), 81; B. Buzan, 'From International System to International Society: Structural Realism and Regime Theory Meet the English School', *International Organization*, vol. 47, no. 3 (1993), 349; and K. Mahbubani, 'The West and the Rest', *The National Interest*, no. 28 (Summer 1992), 5.

Bibliography

Acton, J., *Lectures on Modern History* ed. with an introduction by J. N. Figgis and R. V. Laurence (London: Macmillan, 1950).

Acton, J., *Essays on Freedom and Power* (New York: Meridian Books, 1955).

Anderson, B., *Imagined Communities: Reflections on the Origins and Spread of Nationalism* (London: Verso, 1983).

Ankersmit, F., 'Reply to Professor Zagorin', in *History and Theory*, vol. 29 (1990), 275–96.

Ankersmit, F. R. and Kellner, H. (eds), *A New Philosophy of History* (Chicago: University of Chicago Press, 1995).

Aristotle, *Politics*, trans. J. Warrington (London: Heron Books in arrangement with J. M. Dent, no date).

Arnold, M., *Culture and Anarchy*, with Introduction by J. D. Wilson (Cambridge: Cambridge University Press, 1935).

Aron, R., 'Philosophy of History', *Chambers Encyclopedia*, vol. 7 (1946), 150–5.

Aron, R., *Introduction to the Philosophy of History: An Essay on the Limits of Historical Objectivity*, trans. G. J. Irwin (London: Weidenfeld & Nicolson, 1961).

Aron, R., *Peace and War: A Theory of International Relations*, trans. by R. Howard and A. B. Fox (Garden City, NY: Anchor Books, 1973).

Augustine, Saint, *The City of God*, 2 vols, trans. John Healey, ed. by R. V. G. Tasker, Everyman edition (London: J. M. Dent, 1945).

Ayob, M. (ed.), *The Politics of Islamic Reassertion* (New York: St. Martin's Press – now Palgrave Macmillan, 1981).

Bagby, P., *Culture and History: Prolegomena to the Comparative Study of Civilizations* (London: Longmans, 1958).

Banks, M., 'The Evolution of International Relations Theory' in M. Banks (ed.), *Conflict in World Society: A New Perspective on International Relations* (Brighton: Wheatsheaf Books, 1984).

Bann, S., *The Invention of History: Essays on the Representation of the Past* (Manchester: Manchester University Press, 1990).

Barkan, O. L., 'The Price Revolution of the Sixteenth Century: A Turning Point in the Economic History of the Near East', *International Journal of Middle Eastern Studies* vol. 6 (1975), 3–28.

Barnes, B., *T. S. Kuhn and Social Science* (London: Macmillan – now Palgrave Macmillan, 1982).

Barraclough, G., 'The Larger View of History', *Times Literary Supplement*, no. 2810 (6 January 1956).

Barraclough, G., *History in a Changing World* (Oxford: Basil Blackwell, 1957).

Barraclough, G., 'Universal History' in H. P. R. Finberg (ed.), *Approaches to History: A Symposium* (London: Routledge & Kegan Paul, 1962).

Barraclough, G., *An Introduction to Contemporary History* (Harmondsworth: Pelican Books, 1967).

Beard, C. A., 'That Noble Dream', in F. Stern (ed.), *The Varieties of History: From Voltaire to the Present* (Cleveland: Meridian Books, 1956).

Becker, C. L., *The Heavenly City of the Eighteenth-Century Philosophers* (New Haven: Yale University Press, 1932).
Beeley, B., 'Islam as a Global Political Force', in A. G. McGrew and P. G. Lewis et al. (eds), *Global Politics: Globalization and the Nation-State* (Cambridge: Polity Press, 1992).
Berdyaev, N., *The Meaning of History*, trans. G. Reavey (London: Geoffrey Bles, The Centenary Press, 1936).
Berlin, I., 'History and Theory: The Concept of Scientific History', *History and Theory*, vol. 1 (1961), 1–31.
Berlin, I., *Four Essays on Liberty* (Oxford: Oxford University Press, 1969).
Bernstein, R., *Beyond Objectivism and Relativism: Science, Hermeneutics, and Praxis* (Oxford: Basil Blackwell, 1983).
Berridge, G. R., *International Politics: States, Power and Conflict since 1945*, 2nd edn (London: Harvester Wheatsheaf, 1992).
Bloom, W., *Personal Identity, National Identity and International Relations* (Cambridge: Cambridge University Press, 1990).
Boas, F., 'The Limitations of the Comparative Method of Anthropology', *Science*, vol. 4 (18 December 1896), 901–8.
Bodin, J., *Six Books of the Commonwealth*, abridged and trans. M. Y. Tooley (Oxford: Basil Blackwell, 1955).
Bowle, J., *The Unity of European History: A Political and Cultural Survey* (London: Jonathan Cape, 1948).
Bozeman, A. B., *Politics and Culture in International History* (Princeton: Princeton University Press, 1960).
Bozeman, A., 'The International Order in a Multicultural World', in H. Bull and A. Watson (eds), *The Expansion of International Society* (Oxford: Clarendon, 1984).
Braudel, F., 'History and Social Sciences: The Long Term', *Social Science Information*, vol. 9, no. 1 (1970), 145–74.
Braudel, F., *On History*, trans. S. Matthews (London: Weidenfeld & Nicolson, 1980).
Braudel, F., *A History of Civilizations*, trans. R. Mayne (Harmondsworth: Penguin Books, 1993).
Brown C., *Understanding International Relations* (Basingstoke: Macmillan – now Palgrave Macmillan, 1997).
Buckle, H. T., *Introduction to the History of Civilization in England*, revd edn by J. M. Robertson (London: George Routledge, no date).
Bull, H., 'International Theory: The Case for a Classical Approach', *World Politics*, vol. 18, no. 3 (1966), 361–77.
Bull, H., 'The Theory of International Politics 1919–1969' in Brian Porter (ed.), *The Aberystwyth Papers: International Politics 1919–1969* (London: Oxford University Press, 1972).
Bull, H., *The Anarchical Society: A Study of Order in World Politics* (London: Macmillan, 1977).
Bull, H., 'The Emergence of a Universal International Society' in H. Bull and A. Watson (eds), *The Expansion of International Society* (Oxford: Clarendon, 1984).
Burchill, S., et al., *Theories of International Relations*, 2nd edn (Basingstoke: Palgrave – now Palgrave Macmillan, 2001).
Burckhardt, J., *Reflections on History*, trans. M. D. H. (London: George Allen & Unwin, 1943).

Burckhardt, J., *The Civilization of the Renaissance in Italy*, trans. S. G. C. Middlemore (London: The Phaidon Press, 1944).
Burke, E., *Reflections on the Revolution in France* (London: J. M. Dent, 1910).
Bury, J. B., 'The Science of History' in *Selected Essays of J. B. Bury*, H. Temperley (ed.), (Cambridge: Cambridge University Press, 1930).
Bury, J. B., *The Idea of Progress: An Inquiry into Its Origin and Growth* (London: Macmillan, 1928).
Butterfield, H., *Christianity and History* (London: G. Bell, 1949).
Butterfield, H., *Man on His Past: The Study of the History of Historical Scholarship* (Cambridge: Cambridge University Press, 1955).
Butterfield, H., *The Whig Interpretation of History* (New York: Norton, 1965).
Butterfield, H., 'The Balance of Power' in H. Butterfield and M. Wight (eds), *Diplomatic Investigations* (London: George Allen & Unwin, 1966).
Butterfield, H., 'The New Diplomacy and Historical Diplomacy' in H. Butterfield and M. Wight (eds), *Diplomatic Investigations*.
Butterfield, H., *The Origins of History* (London: Eyre Methuen, 1981).
Buzan, B., 'From International System to International Society: Structural Realism and Regime Theory Meet the English School', *International Organization* vol. 47, no. 3 (1993), 327–52.
Buzan, B. and Little, R., *International Systems in World History: Remaking the Study of International Relations* (Oxford: Oxford University Press, 2000).
Carnoy, M., *The State and Political Theory* (Princeton: Princeton University Press, 1984).
Carr, E. H., *What is History?*, 2nd edn by R. W. Davies (London: Penguin Books, 1987).
Chadwick, O., *The Secularization of European Society in the Nineteenth Century* (Cambridge: Cambridge University Press, 1975).
Childe, V. G., *Man Makes Himself* (London: Watts, 1936).
Childe, V. G., *What Happened in History* (Harmondsworth: Penguin Books, 1942).
Clubb, J. M., 'History as a Social Science,' *International Social Science Journal*, vol. 33, no. 4 (1981), 596–610.
Coles, P., *The Ottoman Impact on Europe* (London: Thames & Hudson, 1968).
Coll, A. R., *The Wisdom of Statecraft: Sir Herbert Butterfield and the Philosophy of International Politics* (Durham: Duke University Press, 1985).
Collingwood, R. G., *The Principles of Art* (London: Oxford University Press, 1938).
Collingwood, R. G., *An Essay on Metaphysics* (Oxford: Clarendon Press, 1940).
Collingwood, R. G., 'The Three Laws of Politics' in *Hobhouse Memorial Lectures, 1941–1950* (London: Oxford University Press, 1952).
Collingwood, R. G., *The Idea of History* (Oxford: Oxford University Press, 1961).
Collingwood, R. G., *The New Leviathan or Man, Society, Civilization and Barbarism*, revd edn, ed. and introduced by D. Boucher (Oxford: Clarendon Press, 1992).
Cox, R. W., 'Thinking about Civilizations', *Review of International Studies*, vol. 26 (2000), 217–34.
Curtis, M. (ed.), *Religion and Politics in the Middle East* (Boulder: Westview Press, 1982).
Davidson, R. H. 'Ottoman Diplomacy and Its Legacy' in L. Carl Brown (ed.), *Imperial Legacy: The Ottoman Imprint on the Balkans and the Middle East* (New York: Columbia University Press, 1996).

Dawisha, A. (ed.), *Islam in Foreign Policy* (Cambridge: Cambridge University Press, 1983).

Dawson, C., *The Dynamics of World History* (London: Sheed & Ward, 1957).

Dehio, L., *The Precarious Balance: Four Centuries of the European Power Struggle* (New York: Alfred A. Knopf, 1962).

Derrida, J., *The Other Heading: Reflections on Today's Europe*, trans. P. Brault and M. B. Naas (Bloomington: Indiana University Press, 1992).

Descartes, R., *The Philosophical Writings of Descartes*, trans. J. Cottingham et al. (Cambridge: Cambridge University Press, 1984).

Dessouki, A. E. (ed.), *Islamic Resurgence in the Arab World* (New York: Praeger, 1982).

Donelan, M. (ed.), *The Reason of States: A Study in International Political Theory* (London: G. Allen & Unwin, 1978).

Donelan, M., *Elements of International Political Theory* (Oxford: Clarendon Press, 1990).

Dore, R., 'Unity and Diversity in World Culture' in Bull and Watson (eds), *The Expansion of International Society* (Oxford: Clarendon Press, 1984).

Dray, W., *Laws and Explanation in History* (London: Oxford University Press, 1957).

Dunn, J., *Western Political Theory in the Face of the Future* (Cambridge: Cambridge University Press, 1979).

Dunne, T., *Inventing International Society: A History of the English School* (Basingstoke: Macmillan – now Palgrave Macmillan, 1998).

Dunne, T., Cox, M. and Booth, K. (eds), *The Eighty Years' Crisis: International Relations 1919–1999* (Cambridge: Cambridge University Press, 1999).

Dyson, K., *The State Tradition in Western Europe* (Oxford: Martin Robertson, 1980).

Eckhardt, W., *Civilizations, Empires and Wars: A Quantitative History of War* (Jefferson, NC and London: McFarland, 1992).

Elias, N., *The History of Manners*, trans. E. Jephcott (New York: Pantheon Books, 1978).

Eliot, T. S., *Notes Towards the Definition of Culture* (London: Faber & Faber, 1948).

Elliot, Sir C., *Turkey in Europe*, 2nd edn (London: Frank Cass, 1965).

Engels, F., 'Feuerbach and End of Classical German Philosophy' in K. Marx and F. Engels (eds), *Selected Works* (London: Lawrence & Wishart, 1968).

Engels, F., 'Origins of Family, Private Property and State' in K. Marx and F. Engels (eds), *Selected Works* (London: Lawrence & Wishart, 1968).

Esposito, J. L., *The Islamic Threat: Myth or Reality?*, 2nd edn (Oxford: Oxford University Press, 1995).

Feyerabend, P. K., *Against Method: Outline of An Anarchistic Theory of Knowledge* (London: New Left Books, 1975).

Forsyth, M., 'The Tradition of International Law' in T. Nardin and D. R. Mapel (eds), *Traditions of International Ethics* (Cambridge: Cambridge University Press, 1992).

Frankfort, H., *The Birth of Civilization in the Near East* (London: Williams & Norgate, 1951).

Freeman, E. A., *Historical Essays*, 4th series (London: Macmillan, 1892).

Fuller, G. E., 'The Emergence of Central Asia', *Foreign Policy*, no. 78 (Spring 1990), 49–67.

Fuller, G. E. and Lesser, I. O. with P. B. Henze and J. F. Brown, *Turkey's New Geopolitics: From the Balkans to Western China* (Boulder: Westview Press, 1993).
Gadamer, H. G., *Truth and Method* (New York: Seabury Press, 1975).
Gaddis, J. L., 'History, Theory and Common Ground', *International Security*, vol. 22, no. 1 (Summer 1997), 75–85.
Gallie, W. B., *Philosophy and Historical Understanding* (London: Chatto & Windus, 1964).
Gardiner, P., *The Nature of Historical Explanation* (London: Oxford University Press, 1961).
Gellner, E., 'The Scientific Status of the Social Sciences', *International Social Science Journal*, vol. 36, no. 102 (1984), 567–86.
Gibbon, E., *The History of the Decline and Fall of the Roman Empire*, 7 vols, ed. by J. B. Bury (London: Methuen, 1925).
Giddens, A., *Studies in Social and Political Theory* (London: Hutchinson, 1977).
Giddens, A., *The Nation State and Violence* (Cambridge: Polity Press, 1985).
Giddens, A., *Social Theory and Modern Sociology* (Cambridge: Polity Press, 1987).
Gierke, O., *Political Theories of the Middle Ages*, trans. F. Maitland (Cambridge: Cambridge University Press, 1900).
Gilbert, R., *The Unequal Treaties* (London: John Murray, 1929).
Gilpin, R., 'The Global Political System' in J. D. B. Miller and R. J. Vincent (eds), *Order and Violence: Hedley Bull and International Relations* (Oxford: Clarendon Press, 1990).
Ginsberg, M., *The Idea of Progress: A Revaluation* (London: Methuen, 1933).
Glenny, M., 'Carnage in Bosnia, for Starters', *New York Times* (29 July 1993), A23.
Gobineau, J. A. de, *The Inequality of Human Races*, trans. A. Collins (London: William Heinemann, 1915).
Gong, G. W., *The Standard of 'Civilization' in International Society* (Oxford: Clarendon Press, 1984).
Gooch, G. P., *History and Historians in the Nineteenth Century*, 2nd edn (London: Longmans, 1952).
Gossman, L., *Towards a Rational Historiography, Transactions of the American Philosophical Society*, vol. 79, pt. 3 (Philadelphia: American Philosophical Society, 1989).
Greene, G., *The End of the Affair* (Harmondsworth: Penguin Books, 1962).
Greene, G., *The Human Factor* (London: Penguin Books, 1978).
Guizot, F., *The History of Civilization in Europe* (London: Cassell, 1911).
Guizot, F., *Historical Essays and Lectures*, ed. and with an Introduction by S. Mellon (Chicago and London: The University of Chicago Press, 1972).
Guthrie, W. K. C., *A History of Greek Philosophy* (Cambridge: Cambridge University Press, 1969).
Hay, D., *Europe: The Emergence of an Idea* (Edinburgh: Edinburgh University Press, 1957).
Hegel, G. W. F., *Lectures on the Philosophy of World History*, ed. by Johannes Hoffmeister, trans. H. B. Nisbet (Cambridge: Cambridge University Press, 1975).
Held, D., *Political Theory and the Modern State: Essays on State, Power and Democracy* (Cambridge: Polity Press, 1989).
Hempel, C. G., *Aspects of Scientific Explanation* (New York: Free Press, 1965).

Herder, J. G. von, *Reflections on the Philosophy of the History of Mankind*, abridged and with an introduction by F. E. Manuel (Chicago and London: University of Chicago Press, 1968).
Herodotus, *The Persian Wars* (Harmondsworth: Penguin Books, 1972).
Herz, J., *International Politics in the Atomic Age* (New York: Columbia University Press, 1959).
Herz, J., *The Nation State and the Crisis of World Politics* (New York: David McKay, 1976).
Hesse, M., *Revolutions and Reconstructions in the Philosophy of Science* (Brighton: Harvester Press, 1980).
Hinsley, F. H., *Power and the Pursuit of Peace* (Cambridge: Cambridge University Press, 1963).
Hinsley, F. H., *Sovereignty*, 2nd edn (Cambridge: Cambridge University Press, 1986).
Hodgson, M. G. S., *The Venture of Islam: Conscience and History in a World Civilization*, 3 vols (Chicago: University of Chicago Press, 1974).
Hodgson, M. G. S., *Rethinking World History: Essays on Europe, Islam and World History*, ed. with an Introduction and Conclusion by E. Burke, III (Cambridge: Cambridge University Press, 1993).
Hoffman, J., 'Hedley Bull's Conception of International Society and the Future of the State', *Leicester University Discussion Papers in Politics*, no. p 92/2 (March 1992).
Hoffman, J., *Beyond the State: An Introductory Critique* (Cambridge: Polity Press, 1995).
Hoffmann, S., 'An American Social Science: International Relations', *Daedalus*, vol. 106, no. 3 (1977), 41–60.
Hollinger, D. A., 'T.S. Kuhn's Theory of Science and Its Implications for History', *American Historical Review*, vol. 78 (1973), 370–93.
Hollinger, D. A., 'The Return of the Prodigal: The Persistence of Historical Knowledge', *American Historical Review*, vol. 94 (1989), 610–21.
Holsti, K. J., *The Dividing Discipline: Hegemony and Diversity in International Relations* (London: George Allen & Unwin, 1985).
Hourani, A., *Europe and the Middle East* (Berkeley: University of California Press, 1980).
Howard, M., 'The Military Factor in European Expansion' in H. Bull and A. Watson (eds), *The Expansion of International Society* (Oxford: Clarendon, 1984).
Hume, D., 'Of the Balance of Power' in M. Wright (ed.), *Theory and Practice of the Balance of Power 1486–1914: Selected European Writings* (London: J. M. Dent, 1975).
Hunt, L. and Revel, J. (eds), *Histories: French Constructions of the Past* (New York: New Press, 1995).
Huntington, S. P., 'The Clash of Civilizations?' *Foreign Affairs*, vol. 72, no. 3 (Summer 1993), 22–49.
Huntington, S. P., *The Clash of Civilizations and the Remaking of the World Order* (New York: Simon & Schuster, 1996).
Hurewitz, J. C., 'Ottoman Diplomacy and the European States System', *The Middle East Journal*, vol. 15 (Spring 1961), 141–52.
Hurewitz, J. C. (ed.), *The Middle East and North Africa in World Politics: A Documentary Record, Vol. 1 European Expansion, 1535–1914*, 2nd edn (New Haven and London: Yale University Press, 1975).

Inalcik, H., 'Imtiyazat', *Encyclopedia of Islam*, 2nd edn (Leiden: E. J. Brill, 1971).
Inalcik, H., 'The Turkish Impact on the Development of Modern Europe' in K. H. Karpat (ed.), *The Ottoman State and Its Place in World History* (Leiden: E. J. Brill, 1974).
Inalcik, H., *The Economic and Social History of the Ottoman Empire* (Cambridge: Cambridge University Press, 1994).
Jansen, G. H., *Militant Islam* (New York: Harper & Row, 1979).
Jenkins, K. (ed.), *Postmodern History Reader* (New York and London: Routledge, 1997).
Jones, F. C., *Extraterritoriality in Japan and the Diplomatic Relations Resulting in Its Abolition 1853–1899*, reprinted edn (New York: AMS Press, 1970).
Joynt, C. B. and Rescher, N., 'The Problem of Uniqueness in History', *History and Theory*, vol. 1 (1961), 151–62.
Kant, I., 'Idea of a Universal History from a Cosmopolitan Point of View,' in P. Gardiner (ed.), *Theories of History* (New York: Free Press, 1959).
Kant, I., *Critique of Pure Reason*, trans. N. K. Smith (London: Macmillan, 1968).
Kaplan, M. A., 'The New Great Debate: Traditionalism vs. Science in International Politics', *World Politics*, vol. 19, no. 1 (1966), 1–20.
Keat, R. and Urry, J., *Social Theory as Science* (London: Routledge & Kegan Paul, 1975).
Kedourie, E., *Nationalism* (London: Hutchinson, 1960).
Kedourie, E., 'Introduction' in *Nationalism in Asia and Africa*, ed., with an introduction by E. Kedourie (London: Frank Cass, 1970).
Keens-Soper, M., 'The Practice of a States-System', in M. Donelan (ed.), *The Reason of States* (London: Routledge & Kegan Paul, 1978).
Kellner, H., *Language and Historical Representation: Getting the Story Crooked* (Madison: University of Wisconsin Press, 1989).
Kennedy, P., *The Rise and Fall of the Great Powers: Economic Change and Military Conflict from 1500 to 2000* (London: Fontana Press, 1989).
Keohane, R. O. and Nye, J. S. (eds) *Transnational Relations and World Politics* (Cambridge, Mass.: Harvard University Press, 1972).
Keohane, R. O. and Nye, J. S., *Power and Interdependence* (Boston: Little, Brown, 1977).
Khaldun, Ibn, *The Muqaddimah: An Introduction to History*, trans. F. Rosenthal, 2nd edn (London: Routledge & Kegan Paul, 1967).
Kipling, R., 'White Man's Burden', *McClure's Magazine*, 12 (February 1899).
Knorr, K. and Rosenau, J. N. (eds), *Contending Approaches to International Politics* (Princeton: Princeton University Press, 1969).
Kohn, H., *The Idea of Nationalism: A Study in Its Origins and Background* (New York: The Macmillan Company, 1944).
The Koran: Interpreted, trans., with an Introduction by A. J. Arberry (Oxford: Oxford University Press, 1983).
Kramer, S. N., *The Sumerians* (Chicago and London: University of Chicago Press, 1963).
Kramnick, I., 'Reflections on Revolution: Definitions and Explanation in Recent Scholarship', *History and Theory*, vol. 11 (1972), 26–63.
Krieger, L., *Time's Reasons: Philosophies of History Old and New* (Chicago and London: University of Chicago Press, 1989).

Kroeber, A. L. and Kluckhohn, C., *Culture: A Critical Review of Concepts and Definitions* (Cambridge, Mass.: Pleabody Museum of American Archeology and Ethnology, 1952).

Kroeber, A. L., 'The Delimitation of Civilizations', *Journal of the History of Ideas*, vol. 14, no. 2 (1953), 264–75.

Kroeber, A. L., *Styles and Civilizations* (Ithaca: Cornell University Press, 1957).

Kuhn, T. S., 'Reflections on My Critics' in I. Lakatos and A. Musgrave (eds), *Criticism and the Growth of Knowledge* (Cambridge: Cambridge University Press, 1970).

Kuhn, T. S., *The Structure of Scientific Revolutions* (Chicago: University of Chicago Press, 1st edn 1962 and 2nd edn 1970).

Kuhn, T. S., 'The Relations between History and History of Science', *Daedalus*, vol. 100, no. 2 (1971), 271–304.

Kuhn, T. S., *The Essential Tension: Selected Studies in Scientific Tradition and Change* (Chicago and London: University of Chicago Press, 1977).

Kuzminski, A., 'Archetypes and Paradigms: History, Politics, and Persons', *History and Theory*, vol. 25, no. 3 (1986), 225–47.

Laue, T. H. von, *Leopold Ranke: The Formative Years* (Princeton: Princeton University Press, 1950).

Leathes, S. (ed.), *The Cambridge Modern History* (Cambridge: Cambridge University Press, 1903).

Lerner, A. J., 'Transcendence of the Nation: National Identity and the Terrain of the Divine', *Millennium: Journal of International Studies* vol. 20, no. 3 (1991), 407–27.

Levinson, M., Butler, M., McGann, J. and Hamilton, P., *Rethinking Historicism* (Oxford: Basil Blackwell, 1989).

Levi-Strauss, C., *Structural Anthropology* (London: Macmillan, 1976).

Lewis, B., *The Emergence of Modern Turkey* (London: Oxford University Press, 1961).

Lewis, B., 'The Roots of Muslim Rage', *The Atlantic Monthly* (September 1990), 47–60.

Lewis, B., 'Rethinking the Middle East', *Foreign Affairs*, vol. 71, no. 4 (1992), 99–119.

Lewis, B., 'The Return of Islam', in his *Islam and the West* (New York and Oxford: Oxford University Press, 1993).

Linklater, A., *Men and Citizens in the Theory of International Relations* (London: Macmillan – now Palgrave Macmillan, 1982).

Little, R., 'The Systems Approach' in S. Smith (ed.), *International Relations: British and American Approaches* (Oxford: Basil Blackwell, 1985).

Lorimer, J., *Institutes of the Law of Nations*, 2 vols (Edinburgh and London: William Blackwood, 1883).

Louis, W. M. R. (ed.), *The Origins of the Second World War: A. J. P. Taylor and Its Critics* (New York: Wiley, 1972).

Lowie, R. H., *The History of Ethnological Theory* (New York: Rinehart & Company, 1937).

Lubbock, J., *The Origin of Civilization and the Primitive Condition of Man* (London: Longmans, Green, 1911).

Lucretius, *The Nature of the Universe*, trans. R. E. Latham (Harmondsworth: Penguin Books, 1951).

McNeill, W. H., *A World History* (London: Oxford University Press, 1967).
McNeill, W. H., 'Civilization', *The Encyclopedia Americana*, international edn (Danbury: Grolier Incorporated, 1982), 1–6.
McNeill, W. H., 'Organizing Concepts for World History', *Review*, vol. 10, no. 2 (1986), 211–29.
McNeill, W. H., *The Global Condition: Conquerors, Catastrophes, and Community* (Princeton: Princeton University Press, 1992).
McNeill, W. H., 'The Changing Shape of World History', *History and Theory, Theme Issue 34, World Historians and Their Critics*, vol. 34, no. 2 (1995), 8–26.
Mahbubani, K., 'The West and the Rest', *The National Interest*, no. 28 (Summer 1992), 3–12.
Mannheim, K., *Ideology and Utopia* (London: Routledge & Kegan Paul, 1936).
Marshall, P. J., and Williams G., *The Great Map of Mankind: British Perceptions of the World in the Age of the Enlightenment* (London: J. M. Dent, 1982).
Marwick, A., *The Nature of History* (London: Macmillan, 1970).
Marx, K. and Engels, F., 'Manifesto of the Communist Party' in K. Marx and F. Engels, *Selected Works* (London: Lawrence & Wishart, 1968).
Mattingly, G., *Renaissance Diplomacy* (Harmondsworth: Penguin Books, 1955).
Meinecke, F., 'Values and Causalities in History' in F. Stern (ed.), *The Varieties of History: From Voltaire to the Present* (Cleveland: Meridian Books, 1956).
Melko, M., *The Nature of Civilizations* (Boston: Porter Sargent Publisher, 1969).
Mill, J. S., 'Civilization', *The London and Westminster Review*, vol. 28, no. 1 (1836), 1–28.
Mill, J. S., *Considerations on Representative Government* (Oxford: Basil Blackwell, 1946).
Modelski, G., *Long Cycles in World Politics* (London: Macmillan – now Palgrave Macmillan, 1987).
Montagu, M. F. A. (ed.), *Toynbee and History: Critical Essays and Reviews* (Boston: Porter Sargent Publisher, 1956).
Montagu, M. W. Lady, *The Turkish Embassy Letters* (London: 1763) in *The Complete Letters of Lady Mary Wortley Montagu*, vol. I, ed., with an Introduction by R. Halsband (London: Oxford University Press, 1965).
Morgenthau, H. J., *Politics among Nations: The Struggle for Power and Peace*, 3rd edn (New York: Alfred A. Knopf, 1960).
Mortimer, E., *Faith and Power: The Politics of Islam* (London: Faber & Faber, 1982).
Naff, T., 'The Ottoman Empire and the European States System' in H. Bull and A. Watson (eds), *The Expansion of International Society* (Oxford: Clarendon Press, 1984).
Nagel, E., *The Structure of Science* (London: Routledge & Kegan Paul, 1961).
Neumann, I. B., *The Uses of the Other: The East in European Identity Formation* (Manchester: Manchester University Press, 1999).
Neumann, I. B. and Welsh, J. M., 'The Other in European Self-Definition: An Addendum to the Literature on International Society', *Review of International Studies*, vol. 17 (1991), 327–48.
The New English Bible: With the Apocrypha (London: Oxford University Press and Cambridge University Press, 1970).
Nietzsche, F., *Thus Spake Zarathustra*, trans. A. Tille, Everyman edn (London: J. M. Dent, 1933).

Nietzsche, F., *Twilight of the Idols*, trans. R. J. Hollingdale (London: Penguin Books, 1968).
Nietzsche, F., *Will to Power*, trans. W. Kauffmann and R. J. Hollingdale (London: Weidenfeld & Nicolson, 1968).
Nietzsche, F., *Ecce Homo: How One Becomes What* One *Is*, trans. R. J. Hollingdale (London: Penguin Books, 1969).
Nietzsche, F., *On the Genealogy of Morals*, trans. W. Kauffmann and R. J. Hollingdale (New York: Vintage Books, 1969).
Noone, J. B., Jr, 'The Philosophy of History: A Prolegomenon to Political Philosophy', *The Review of Politics*, vol. 23, no. 4 (1961), 472–89.
Northedge, F. S., *The International Political System* (London: Faber & Faber, 1976).
Northedge, F. S. and Grieve, M. J., *A Hundred Years of International Relations* (London: Duckworth, 1971).
Northrop, F. S. C., *The Meeting of East and West: An Inquiry Concerning World Understanding* (New York: The Macmillan Company, 1946).
Novick, P., *That Noble Dream: The 'Objectivity Question' and the American Historical Profession* (New York: Cambridge University Press, 1988).
Oakeshott, M., *Rationalism in Politics and Other Essays* (London: Methuen, 1962).
Oakeshott, M. 'Character of a Modern European State', *On Human Conduct* (Oxford: Clarendon Press, 1975).
Oakeshott, M., *On History and Other Essays* (Oxford: Basil Blackwell, 1983).
Oaks, A. and Mowat, R. B., *The Great European Treaties of the Nineteenth Century* (Oxford: Oxford University Press, 1918).
Olson, W. C., 'The Growth of a Discipline' in Brian Porter (ed.), *The Aberystwyth Papers: International Politics 1919–1969* (London: Oxford University Press, 1972).
Ortega y Gasset, J., *An Interpretation of Universal History*, trans. M. Adams (New York and London: W. W. Norton, 1973).
Osiander, A., *The States System of Europe 1640–1990: Peacemaking and the Conditions of International Stability* (Oxford: Clarendon Press, 1994).
Osiander, A., 'Sovereignty, International Relations, and the Westphalian Myth', *International Organization*, vol. 55 (2001), 251–87.
Panikkar, K. M., *Asia and Western Dominance* (London: George Allen & Unwin, 1953).
Peyrefitte, A., *The Collision of Two Civilizations: The British Expedition to China 1792–4*, trans. J. Rothschild (London: Harvill, 1993).
Pipes, D., *In the Path of God: Islam and Political Power* (New York: Basic Books, 1983).
Piscatori, J., 'Islam in International Order', in H. Bull and A. Watson (eds), *The Expansion of International Society* (Oxford: Clarendon Press, 1984).
Plato, *Republic*, trans. A. D. Lindsay (London: Heron Books, in arrangement with J. M. Dent, no date).
Popper, K. R., *Logic of Scientific Discovery* (New York: Basic Books, 1959).
Popper, K. R., 'Normal Science and Its Dangers' in I. Lakatos and A. Musgrave (eds), *Criticism and the Growth of Scientific Knowledge* (Cambridge: Cambridge University Press, 1970).
Popper, K. R., *The Poverty of Historicism* (London: Ark Paperbacks, 1986).
Quigley, C., *The Evolution of Civilizations: An Introduction to Historical Analysis* (New York: The Macmillan Company, 1961).

Raglan, Lord, *How Came Civilization* (London: Methuen, 1939).

Ranke, L. von, *Universal History: The Oldest Historical Groups of Nations and the Greeks*, trans. and ed. by G. W. Prothero (London: Kegan Paul, Trench, & Co., 1884).

Ranke, L. von, 'A Dialogue on Politics' in T. H. von Laue (ed.), *Leopold Ranke: The Formative Years* (Princeton: Princeton University Press, 1950).

Ranke, L. von, *The Secret of World History: Selected Writings on the Art and Science of History*, ed. with trans. Roger Wines (New York: Fordham University Press, 1981).

Redfield, R., *The Primitive World and Its Transformations* (Ithaca: Cornell University Press, 1953).

Ritter, G., 'Scientific History, Contemporary History, and Political Science', *History and Theory*, vol. 1 (1961), 261–79.

Robins, P., 'Between Sentiment and Self-Interest: Turkey's Policy Toward Azerbaijan and the Central Asian States', *Middle East Journal*, vol. 47, no. 4 (1993), 593–610.

Rodinson, M., *Europe and the Mystique of Islam* (London: I. B. Tauris, 1987).

Rousseau, J. J., *The Social Contract and Discourses*, trans. G. D. H. Cole (London: J. M. Dent, 1913).

Russell, B., *History of Western Philosophy and Its Connections with Political and Social Circumstances from the Earliest Times to the Present Day* (London: Readers Union with Allen & Unwin, 1954).

Sanders, N. K., *The Epic of Gilgamesh* (Harmondsworth: Penguin Books, 1960).

Savigear, P., 'International Relations and the Philosophy of History' in M. Donelan (ed.), *The Reason of States* (London: George Allen & Unwin, 1978).

Schmidt, B. C., *The Political Discourse of Anarchy: A Disciplinary History of International Relations* (Albany: State University of New York Press, 1998).

Seeley, J., *Introduction to Political Science* (London: Macmillan, 1896).

Shapcott, R., 'Conversation and Coexistence: Gadamer and the Interpretation of International Society', *Millennium: Journal of International Studies*, vol. 23, no. 1 (1994), 57–83.

Shapere, D., 'The Structure of Scientific Revolutions', *Philosophical Review*, vol. 73 (1964), 383–94.

Sked, A., 'The Study of International Relations: A Historian's View', *Millennium: Journal of International Studies*, vol. 16, no. 2 (1987), 251–62.

Skinner, Q., 'Meaning and Understanding in the History of Ideas', *History and Theory*, vol. 8 (1969), 3–53.

Smith, A. D., *National Identity* (London: Penguin Books, 1991).

Smith, S., 'Paradigm Dominance in International Relations: The Development of International Relations as a Social Science', *Millennium: Journal of International Studies*, vol. 16, no. 2 (1987), 189–206.

Snow, C. P., *The Two Cultures and the Scientific Revolution*, with Introduction by S. Collini (Cambridge: Cambridge University Press, 1993).

Sorokin, P. A., *Modern Historical and Social Philosophies* (New York: Dover Publications, 1963).

Sousa, N., *The Capitulatory Regime of Turkey: Its History, Origin and Nature* (Baltimore: Johns Hopkins Press, 1933).

Southern, R. W., *The Making of the Middle Ages* (London: Hutchinson University Library, 1953).

Spalding, H. N., *Civilization in East and West: An Introduction to the Study of Human Progress* (London: Oxford University Press, 1930).
Spengler, O., *The Decline of the West*, 2 vols, trans. C. F. Atkinson (New York: Alfred A. Knopf, 1926–1928).
Starn, R., 'Historians and "Crisis"', *Past and Present*, no. 52 (August 1971), 3–22.
Stavrianos, L. S., 'The Teaching of World History', *The Journal of Modern History*, vol. 31 (1959), 110–17.
Strauss, L., 'Political Philosophy and History', *Journal of the History of Ideas*, vol. 10, no. 1 (1949), 30–50.
Suganami, H., *The Domestic Analogy and World Order Proposals* (Cambridge: Cambridge University Press, 1989).
Taylor, A. J. P., *The Origins of the Second World War* (London: Hamish Hamilton, 1961).
Temperley, H., 'The Treaty of Paris of 1856 and Its Execution', *Journal of Modern History*, vol. 4 (1932), 387–414, 523–43.
Thucydides, *History of the Peloponnesian War*, trans. R. Crawley (London: J. M. Dent, 1910).
Toynbee, A. J., *Civilization on Trial* (London: Oxford University Press, 1948).
Toynbee, A. J., *An Historian's Approach to Religion* (London: Oxford University Press, 1956).
Toynbee, A. J., *A Study of History*, 12 vols (London: Oxford University Press, 1934–1961).
Toynbee, A. J., *Hannibal's Legacy: The Hannibalic War's Effects on Roman Life*, 2 vols (London: Oxford University Press, 1965).
Trevelyan, G. M., *History of England* (London: Longmans, 1943).
Tylor, E. B., *Primitive Culture: Researches into the Development of Mythology, Philosophy, Religion, Language, Art, and Custom* (First published 1871; New York: Brentanos Publishers, 1903).
Valéry, P., *History and Politics*, trans. D. Folliot and J. Mathews (London: Routledge & Kegan Paul, 1962).
Vaughan, D. M., *Europe and the Turk: A Pattern of Alliances, 1350–1700* (Liverpool: Liverpool University Press, 1954).
Veeser, H. A., (ed.), *The New Historicism* (New York and London: Routledge, 1989).
Vico, G., *The New Science*, trans. T. B. Gergen and M. H. Fisch (Ithaca: Cornell University Press, 1948).
Vincent, A., *Theories of the State* (Oxford: Basil Blackwell, 1987).
Vincent, R. J., 'Race in International Relations', *International Affairs*, vol. 58, no. 4 (1982), 658–70.
Vincent, R. J., 'Racial Equality' in H. Bull and A. Watson (eds), *The Expansion of International Society* (Oxford: Clarendon, 1984).
Voltaire, F., *The Age of Louis XIV* (London: J. M. Dent, 1935).
Walker, R. B. J., *Inside/Outside: International Relations as Political Theory* (Cambridge: Cambridge University Press, 1993).
Wallerstein, I., *The Modern World-System* (New York: Academic Press, 1974).
Wallerstein, I., *The Capitalist World Economy* (Cambridge: Cambridge University Press, 1979).
Wallerstein, I., 'World System Analysis' in A. Giddens and J. Turner (eds), *Social Theory Today* (Cambridge: Polity Press, 1987).

Wallerstein, I., 'The National and Universal: Can There be such a Thing as World Culture?' in *Geopolitics and Geoculture: Essays on the Changing World-System* (Cambridge: Cambridge University Press, 1991).

Walsh, W. H., *An Introduction to Philosophy of History*, 3rd edn (London: Hutchinson, 1967).

Watson, A., 'European International Society and Its Expansion' in H. Bull and A. Watson (eds), *The Expansion of International Society* (Oxford: Clarendon, 1984).

Watson, A., 'Hedley Bull, States Systems and International Societies', *Review of International Studies*, vol. 13 (1987), 147–53.

Watson, A., *The Evolution of International Society: A Comparative Historical Analysis* (London: Routledge, 1992).

Weber, M., 'Politics as a Vocation' in *From Max Weber: Essays in Sociology*, trans. and ed. by H. H. Gerth and C. Wright Mills (New York: Oxford University Press, 1946).

Webster, C. K., 'The Study of International History', *History*, vol. 17 (1933), 97–114.

Wells, H. G., *The Outline of History: Being a Plain History of Life and Mankind* (London: Cassell, 1920).

White, W. H., 'The Politics of Historical Interpretation: Discipline and De-Sublimation', *Critical Inquiry*, vol. 9 (1982), 113–37.

Whitrow, G. J., *What Is Time?* (London: Thames & Hudson, 1972).

Whitrow, G. J., *Time in History: Views of Time from Prehistory to the Present Day* (Oxford: Oxford University Press, 1989).

Whyte, A. F., *China and Foreign Powers* (London: Milford, 1927).

Wight, M., *Systems of States*, ed. with an Introduction by H. Bull (Leicester: Leicester University Press, 1977).

Winch, P. G., *The Idea of a Social Science and Its Relation to Philosophy* (London: Routledge & Kegan Paul, 1958).

Winrow, G., *Where East Meets West: Turkey and The Balkans* (London: Institute for European Defence and Strategic Studies, 1993).

Wittgenstein, L., *Philosophical Investigations*, trans. G. E. M. Anscombe (Oxford: Basil Blackwell, 1978).

Wood, H. McKinnon, 'The Treaty of Paris and Turkey's Status in International Law', *American Journal of International Law* (1943), 264–74.

Wright, A. F., 'The Study of Chinese Civilization', *Journal of the History of Ideas*, vol. 21 (1960), 233–55.

Wright, Q., *A Study of War*, 2nd edn (Chicago: Chicago University Press, 1965).

Yurdusev, A. N., '"Level of Analysis" and "Unit of Analysis": A Case for Distinction', *Millennium: Journal of International Studies*, vol. 22, no. 1 (1993), 77–88.

Yurdusev, A. N., 'The Concept of International System as a Unit of Analysis', *METU Studies in Development* (Ankara), vol. 21, no. 1 (1994), 143–74.

Yurdusev, A. N., 'Turkey and Europe: The Other in Identity Formation', *Zeitschrift fur Turkeistudien*, 13. Jahrgang 2000, Heft 1, 85–94.

Zaehner, R. C., *Dawn and Twilight of Zoroastrianism* (London: Weidenfeld & Nicolson, 1961).

Zagorin, P., 'Historiography and Postmodernism: Reconsiderations', *History and Theory*, vol. 29 (1990), 263–74.

Zammito, J. H., 'Are We Being Theoretical Yet? The New Historicism, The New Philosophy of History and "Practicing Historians"', *Journal of Modern History*, vol. 65, no. 4 (1993), 783–814.
Zolberg, A. R., 'Origins of the Modern World System: A Missing Link', *World Politics*, vol. 23, no. 2 (1981), 253–81.

Index